Discovery-Based Learning in the Life Sciences

Discovery-Based Learning in the Life Sciences

Kathleen M. Susman

WILEY Blackwell

Published by John Wiley & Sons, Inc., Hoboken, New Jersey
Published simultaneously in Canada

For general information on our other products and services or for technical support, please contact our Customer Care Department within the United States at (800) 762-2974, outside the United States at (317) 572-3993 or fax (317) 572-4002.

Wiley also publishes its books in a variety of electronic formats. Some content that appears in print may not be available in electronic formats. For more information about Wiley products, visit our web site at www.wiley.com.

Library of Congress Cataloging-in-Publication Data:

Susman, Kathleen M., author.
  Discovery-based learning in the life sciences / Kathleen M. Susman.
      pages cm
  Includes index.
  ISBN 978-1-118-90756-6 (pbk.)
  1. Life sciences–Study and teaching. 2. Learning by discovery. I. Title.
  QH315.S87 2015
  570.76–dc23
                                    2015007224

Printed in Singapore by C.O.S. Printers Pte Ltd
10 9 8 7 6 5 4 3 2 1

1 2015

Dedication

To David, Patrick, Melanie and Daniel, the loves of my life

Contents

Acknowledgments

I hope that you will find some ideas in this book that you can use in your own teaching. I thank you for giving these ideas a try. In the more than 20 years since I designed and taught my first solo course, I have benefited from the ideas and encouragement of my colleagues. My hope is that this book will serve as a source of ideas and encouragement for you.

This book is built from the hard work and creativity of many individuals. I am so grateful to each for generously providing ideas and insights into new approaches to teaching. I thank all the individuals whose work is published online and in education journals. Their generosity of spirit, providing their work to other educators freely, forges a supportive community of teacher/scholars that undergirds this revolution in teaching that is underway. My own colleagues at Vassar College, both current and former, are a true inspiration: John H. Long, Jr., Margaret Ronsheim, Sarah Kozloff, Mark A. Schlessman, A. Marshall Pregnall, Jodi Schwarz, Janet Gray, David Esteban, David Jemiolo, J. William Straus, Leathem Mehaffey, Ann Mehaffey, Mary Ellen Czesak, Elizabeth Collins, Nancy Pokrywka, Lynn Christenson, Carol Christensen, Molly McGlennen, Susan Zlotnick, Jenni Kennell, Megan Gall, Kelli Duncan, Erica Crespi, Jeremy Davis, Cynthia Damer, E. Pinina Norrod, Robert Suter, Richard Hemmes, Robert Fritz, Jennifer Turner Waldo, and Marie Pizzorno. Through conversations at the Xerox machine and countless more outside of classrooms and laboratories, through team-teaching and curriculum committee meetings, I owe a debt of gratitude to each and every one. Through them I have learned to teach and to become a biologist. Thank you also to colleagues at *Caenorhabditis elegans* meetings and the Faculty for Undergraduate Neuroscience.

None of us teaches in a vacuum. All of what we do rests on the shoulders of those who came before us and who stand along side us. Thank you to inspiring teachers from high school and from college, graduate school and postdoctoral studies. I also owe a debt of gratitude to all the students who were guinea pigs for many of these ideas. You are the reason we do what we do.

I could never have put fingers to keyboard to craft this book without the support of my family: Dan, Melanie, Patrick, and David. Thanks also to my parents, brother, sister, mother-in-law and father-in-law for helpful advice and encouragement. Even my dog Pippin put up with me sitting for hours on end, laptop in lap. He might actually have benefitted the most from the closeness that he always craves being more available from all the hours spent sitting with him cuddled up against my thigh while I wrote. Everyone else tolerated my absences, my writing during swim practices and gymnastics classes, during family game nights by the fire and in the early mornings before the bus. I love you all and dedicate this book to you.

1 The New Life Sciences

Biology in the twenty-first century is not your grandmother's or even your mother's biology. Disappearing are the highly specialized silos of biological knowledge – biochemistry, molecular biology, and cell biology – and the reductionistic imperative to start at the bottom and work our way up the organizational chart of life. Gone is the sense that to understand any biological phenomenon one must reduce it to its most basic level. Biologists are through with the need to view everything as an extreme close-up. From that vantage, life is featureless, static, and dull.

Our new biology is a complex, dynamic web of interactions. A nonlinear connectome. Sure the building blocks, the units of life, are still there. But the whole is more than the sum of the parts. Life is complicated, ever changing, and intriguing. Biologists need now be nimble, willing to question long-held beliefs about how organisms are related, how they change. Here's a quick example – the species concept. My generation learned that a species is a group of populations of related organisms that can potentially interbreed to produce viable offspring. Just a couple of questions will illustrate a few of the problems with this simple definition.

- What about the vast array of organisms that reproduce asexually?
- What about the organisms that look very different but can interbreed under some circumstances, but that normally don't?
- Or, where members of the same species are so different that they do not or cannot interbreed? Such as, for example, a Great Dane and a Chihuahua?

Newer species concepts include genetic, morphological, and ecological factors. Nowadays, phylogenetic trees, once thought to be a static tree of life (Figure 1.1) are instead considered working models, subject to revision (Figure 1.2) as we learn more.

Discovery-Based Learning in the Life Sciences, First Edition. Kathleen M. Susman.
© 2015 John Wiley & Sons, Inc. Published 2015 by John Wiley & Sons, Inc.

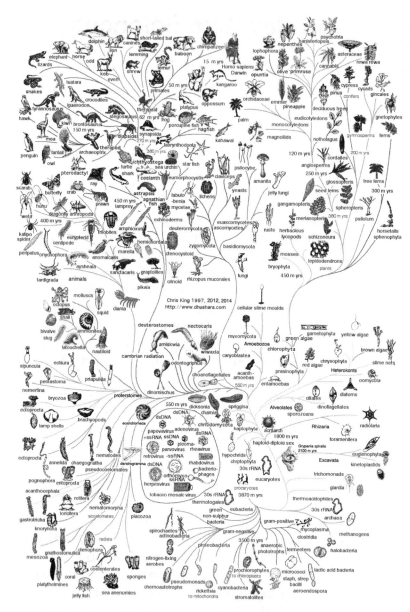

Figure 1.1   A typical tree of life based on morphological characters. (Courtesy: Chris King. www.dhushara.com/book/unraveltree/unravel.htm.)

The Challenges We Face in Teaching the New Biology

New technologies and paradigm-shifting approaches have dramatically transformed the life sciences over the past 20 years. Initially, college-level teaching accommodated this break-neck pace of discovery by adopting

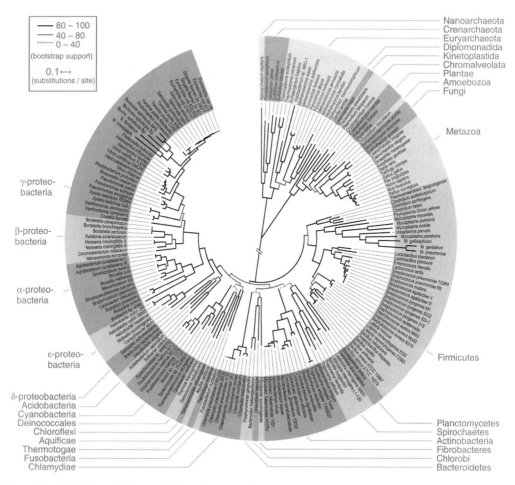

Figure 1.2 A phylogenetic tree of life no longer really a tree, more like a swirl, as in this representation of evolutionary relationships, based on a genomic study of the rRNA of 3,000 species by David Hillis, Derrick Zwicki and Robin Gutell from the University of Texas. (From Ciccarelli et al. (2006). Reprinted with permission from AAAS.) (*See insert for color representation of this figure.*)

encyclopedic textbooks and streamlining lectures with full color slides instead of writing on the chalkboard. College instructors now fret about how to "cover" all the new material while continuing to teach the "classic" fundamental content from the 1980s and earlier. After all, the life sciences that we professors experienced began with the structure of DNA and the Central Dogma of DNA to RNA to protein and moved up levels of organization to consider Hardy–Weinberg's ideas of population-level evolutionary change and broader ecological principles such as succession. How could we not continue to teach those classic ideas and the elegant experiments that led to them?

At the same time, a growing field of pedagogical scholarship focuses on how students learn and retain information. The content-rich, lecture-style classroom experience, while still widely used, has fallen out of favor, making way for student-centered discussion and activities-based teaching strategies that, along with some lecture format, enhance the overall learning experience for our students.

Many professors of college introductory courses in the life sciences struggle to both teach all the new content (while retaining the "classics") and incorporate new teaching and learning strategies.

Static for at least 50 years, introductory biology courses continue to be organized as the lecture period along with a weekly multihour laboratory component. In particular, the laboratory component has remained virtually unchanged for many college introductory biology courses. A glance at the commercially available introductory biology laboratory manuals on the market today reveals the same structure as those manuals published more than 20 years ago. Textbooks have also changed little except for vastly more content and much better illustrations and accompanying multimedia materials such as animations and videos. Authors of the newest textbooks have added question boxes and experimental research sidebars to enliven and provide some inquiry-based focus, but the same basic organization, the same march through content beginning with the molecules of life and ending with ecosystems, provides an overwhelming impression of biology as "a bunch of facts."

There is too much information to "cover." The move away from lecture to more student-centered, active approaches makes it even more difficult to "cover" content. The content overload has trickled down to high school and even middle school science classes. Children begin learning the language of science, but what's left by the time they get to college is just word recognition. In high school, they memorize lists of terms: "mitochondrion = fuel source of the cell;" with little depth about how that fuel is produced. Children (and many teachers) believe that hearing or memorizing terms and concepts equals learning. Been there, done that, they think. A quick quiz at the beginning of a college introductory biology course would soon reveal how little was remembered from high school, how little is understood. Like a language, Biology needs to be practiced; terms need to be used, spoken, written, and formed into new combinations.

And forget about laboratories and the process of science. One-time experiences and demonstrations do not teach students how science is done. If anything, these brief experiences convey false impressions that doing science is all about confirming what we already know. Students come away from these experiences believing that the ideas are "proven" by doing experiments. They don't see or experience the discovery process, the creative side of science. These misperceptions and attitudes then enter the introductory biology classroom at the college level, not to mention our society's views of science and scientists. Dimly remembered bits and pieces of facts and vocabulary words engrain falsehoods about how science is done. We have our work cut out for us, to be sure.

Visions of Change

Many college students take a course in introductory biology or life science as a requirement to satisfy distributional or science general education requirements. This means that introductory biology is a great opportunity to make a real difference in how science, life sciences in particular, is perceived. It's probably safe to say that most of us teaching introductory life sciences would agree that college-educated citizens and future policy-makers and leaders need a firm understanding of the major concepts of biology. Climate change and its impacts are routinely in the news and are already affecting communities worldwide. Antibiotic-resistant superbugs and the global spread of new infectious diseases make the headlines. We all worry about getting cancer or want to age without Alzheimer's disease. Teaching biology in a way that promotes engagement rather than dismissal is thus a critical mission for college educators. Scientists and science education policy makers are urging life sciences educators to transform their teaching practices to better represent this new life science and to dispel the misconceptions about science. Because introductory and intermediate-level life sciences courses are the major training grounds for the vast majority of our educated citizens to be exposed to science in any form, there's a lot riding on them. We need to reinvent how we teach these courses.

Not all scientists agree that we need to change how we are teaching introductory biology, however. Some reason that scientists have been learning this way, through lectures that march up level of organization and through laboratories that introduce a new technique virtually every week, successfully for decades, so why do we need to change? "It worked for me. I'm a successful scientist and college professor. Why change?"

Perhaps the biggest reason urging us to change our approach to the teaching of life sciences is that it *hasn't* worked to increase general science literacy. In fact, the plod through a 1200-page textbook coupled with content-filled lectures and fact-laden examinations has been one of the biggest turn offs for most students for decades. Unfortunately, by the time most students get through middle school, they have turned off to science.

Then, after high school biology, taught as facts and concepts to be memorized, most students close their minds to learning about biology. "Biology is boring." These high-school students graduate and either go to college or enter the workplace, and many put their biology experiences behind them, never again to try to learn more about the life sciences, except what they glean from the evening news or the *Discovery Channel*.

College-level introductory biology as it is currently taught at most institutions does not succeed in re-opening these minds to the wonders of the living world, nor does it succeed in developing science literacy or appreciation. Those who go to college and take an introductory biology course often have their high school experiences further validated by the content and memorization-focused college lecture class structure and the multiple choice and scantron examinations. Laboratory experiences are little more than a disjointed series of different procedures, where students follow numbered lists of steps punctuated with questions: "What was the purpose of adding heat-killed peas to a tube? What does this experiment tell you about the influence of temperature on oxygen consumption during cellular respiration?" (Vodopich and Moore 10th Ed.; see Further Reading). The career biologists, physicians, and chemists simply persisted through those courses, despite the lackluster, fact-packed, memorization-focused approaches in class and the "cookbook"-style, procedure-focused laboratory sessions. The rest leave biology and science in droves, never looking back. We need to fundamentally change how we teach life sciences at the undergraduate level if we want our citizens and leaders to better appreciate the importance of scientific ways of thinking and to better understand the planet we live on and our effects on its inhabitants, including us.

Need for Structural Change

All students benefit from learning and understanding the scientific process. Because science is a part of all the major issues of modern life, appreciating how the scientific process works helps us critically evaluate what we read in the newspaper about the benefits or hazards of the latest fad diet or forming an opinion about the dangers and benefits of hydraulic fracturing. Understanding the creative process involved in discovering new knowledge about the world helps all citizens navigate the information-overload age, to assess the value or accuracy of the newest scientific claims or to appreciate the benefits to society of basic research and the importance of national funding of basic research. Applying skills of scientific literature searching and reading to the process of gathering evidence to understand questions in real life, such as medical care, financial management, and the adoption of new technologies, helps and empowers all of us to make informed decisions. These skills are the tool-kit for self-discovery. Engaging all students in science should therefore be a critical goal of all colleges and universities.

Is it fair to place the responsibility for the science literacy of our college graduates on a single introductory science course such as introductory biology? Of course not. In an ideal world, our college students would take more than one science course and would ideally get beyond introductory material into more intermediate or advanced-level thinking in a scientific discipline. Unfortunately, the vast majority of students will not penetrate a science curriculum more deeply than that first course. But we can certainly improve things with a single course. One outcome should be to inspire college students to take *more than one* science course. If our introductory curricula are designed to get students interested in learning more, we will be on the road to real change.

For over 20 years, STEM (Science, Technology, Engineering, and Mathematics) policy makers and educators have been issuing reports and increasingly urgent calls to revamp how science and math are taught at all levels of education. Active learning, relevance of instructional materials to the daily lives of students, and enhanced excitement of curricular materials are frequently considered the appropriate mechanisms for building a more science-savvy and more science-welcoming populace. These changes are difficult to implement into large classes with limited budgets and resources and by educators not rewarded for the time and effort spent in redesigning courses. Furthermore, science professors are trained in research techniques, not in teaching techniques. Most perpetuate the teaching methods that they were exposed to as college and graduate students, maybe at best incorporating a few new strategies into an existing format and structure that is often viewed as immovably institutionalized.

While making wholesale changes to your introductory biology curriculum isn't easy, I will show in this book a distillation of steps you can take to modify these long-held structures and gradually transform how you package and deliver key biological concepts. By just rearranging the order of your laboratory exercises, your students can more fully appreciate the process of discovery, to see the relevance of the laboratory experiences to their personal lives. By introducing a conceptual topic in your classroom sessions and continuing that topic in the laboratory, you will enhance your students' learning and retention of key biological concepts. You might even inspire them to continue taking life sciences courses and help them appreciate the importance of science to their future lives.

To enact lasting change and to incorporate successful strategies to improve the science literacy of our college graduates will require small changes that are driven first by individuals, then by departments, and finally by institutions. Chapter 3 presents some ideas and strategies for reorganizing an introductory biology laboratory course into conceptual topics that will get you on your way. The reorganization, because it's conceptual, still works within your institutional framework and uses few additional resources. Once you've made these organizational changes, you'll be ready for the more substantial structural changes I suggest later in the book.

Conceptual Organization of Introductory Biology

When looked at through the lens of current biological research, five over-arching concepts or themes percolate throughout all levels of biological organization. Beginning with the 1985 "Benchmarks for Science Literacy" proposed by the AAAS and reinforced by subsequent nationally sponsored reports, the following core concepts have been identified in "Vision and Change 2011" (see Further Reading):

(1) Evolution

   The processes of inheritance, adaptation, and change yield the tremendous biological diversity that evolved on planet Earth over its four-billion-year history. These processes operate at every level of biological organization, from molecular genetics to physiology and biochemistry to organisms, ecology, and paleontology. Mechanisms of evolutionary change such as natural selection, genetic drift, mutation, and gene flow can be taught by considering relevant topics such as antibiotic resistance, artificial selection events such as agricultural practice, pesticide resistance, and domestication of pets and livestock. Laboratory sequences and modules such as those described in Chapters 3, 5, and 6 can allow students to discover these mechanisms for themselves.

(2) Structure and function

   All biological systems are built from interacting units, from macro-molecules to cells to tissues to organisms. These functional units operate similarly for all organisms. Even organisms as dissimilar as bacteria and palm trees, when viewed in terms of structural/functional units, share many characteristics and processes at many levels of organization. Understanding how smaller functional units such as cells interact to form larger functional units such as tissues, organs, organisms, and populations reveals key insights into how living processes work and are organized. Laboratory sequences such as those described in this book reinforce ideas of structure/function relationships that can also inform the discovery of these ideas.

(3) Information flow, exchange, and storage

What is the nature of biological information? Information exists within the structure of molecules such as proteins and DNA. Information is also delivered and exchanged in biochemical signaling pathways, in the expression patterns of genes and gene regulatory networks, in the process of development of multicellular organisms from zygote to functioning adult. Populations of interacting organisms together form species that share and exchange information in the form of genes, signaling compounds such as pheromones, sensory communication signals, behaviors, and social structures. Ecosystems of interacting species organize and exchange information about genetic and evolutionary relatedness, nutrient flow, and cycling. Laboratory experiences such as those described in Chapter 5 allow students to explore these interconnections through discovery-based projects.

(4) Pathways and transformations of matter

All living things use and transform energy. From cells using energy currency in the form of ATP hydrolysis to power cytoskeletal dynamics, to ecosystems cycling nutrients from soil and air, understanding how living systems make, use, store, and transform energy links biochemistry with the laws of thermodynamics. Processes such as homeostasis, membrane dynamics, and metabolic pathways operate within the confines of physics and chemistry. Topics of current global interest such as bioremediation of oil spills or the production of green energies provide relevant examples of these principles. Laboratory modules, described in Chapter 3, which link energy transformation such as photosynthesis with extraction of plant photopigments and plant growth help students discover and reinforce these interconnected ideas.

(5) Systems

Complexity arises within subcellular organelles in interconnected metabolic pathways. Complexity exists within interacting populations in an ecosystem that exchanges nutrients and information. Taking a systems approach to topics such as cellular respiration, cell signaling, development, and population ecology underscores the interrelatedness and interconnectedness of living systems. Laboratory modules such as those described in Chapter 3 that emphasize quantitative approaches such as simulations of population growth or bioinformatic analysis of gene regulatory networks help students discover how dynamic interactions between and among levels of biological organization give rise to emergent functional properties such as the development of multicellular organisms.

This book focuses on how to organize and implement laboratory modules that explicitly address these conceptual areas. In addition, the modules emphasize the processes by which scientists explore these conceptual areas. Each discovery-based laboratory module provides students with practice developing a question of interest into a testable hypothesis. With practice designing experiments and refining experiments, with trouble-shooting and extending an experimental project, the emphasis is on *the process of experimentation*, rather than on the acquisition of technical skills. That acquisition will just happen along with way.

Learning and Mastering

We learn by doing. We improve by making mistakes and trying again. We master through practice. We learn through feeling personally motivated, by *wanting* to learn. These ideas about learning are not new. Think about how you learned to master anything, from riding a bicycle to speaking a new language to learning to read. Why should learning science be any different?

Despite ongoing debate among college educators, there is growing support for the use of student-centered, inquiry-driven, problem-based learning in college-level science courses, particularly at the introductory and intermediate levels. There are many online resources and articles published in education journals such as *Cell Biology Life Education* describing these types of pedagogical approaches to learning science. A key aspect of these strategies that likely plays a dominant role in their success is the emphasis on active student participation. Rather than simply sitting passively day after day in a lecture hall as the information is presented in the form of slides, animations, and an entertaining professor, students grapple with questions and problems and work in small groups to arrive at an answer or articulation of a concept or issue in class. There is immediate feedback, from the looks of comprehension on the faces of student peers during a discussion or problem-solving session, to the satisfaction of having successfully solved a problem, to the thrill of understanding something new after doing it or seeing it up close and personal. The following chapter considers some of the strategies that can be used in introductory biology classes and laboratories, no matter the size. These strategies all require that the students work more actively with the subject matter. Students practice the language, practice applying the concept, work with the material, and learn from their mistakes. If done effectively, as described in the following chapter, these strategies are all natural ways that we tend to learn just about anything.

The laboratory portion of most introductory science courses, as emphasized in Chapters 3 and 5, is a key learning environment that is not utilized nearly effectively enough. Even those laboratories that have been revised to incorporate inquiry-driven experiences are often little more than guided exercises that still give inadequate practice and inadequate active work with the material. An essential element missing from most laboratory sequences is the practice, the learning by re-doing. Because laboratories are in essence skills-learning opportunities, it is critical to organize them that way. Think about the first time you learned a complex skill. Let's take an example such as learning to use a digital camera. If you are provided an instruction manual and told to read it and make an outline of the contents, is that enough to learn how to take a good digital photograph? Then, the next day, you are asked to use your outline and the instruction manual to take a photograph. You take the photograph, transfer the digital image to your computer, and then make a printout of your photograph. If that's the only time you use the camera, but then move on to a different technique the next week, will you have learned digital photography?

In reality, we take a whole bunch of digital photographs. We try all the different settings, we take a lot of really poor quality photos, and after many trials, we begin to understand better how to set up the camera to yield a certain type of photograph, such as a landscape photo on a sunny day or a portrait taken without backlighting. After much practice, we get pretty good at it. We also begin to understand more about lighting and some of the more conceptual aspects of photography.

How can we design our introductory laboratory experiences to include more elements of practice? What do we want our students to practice? Do we want them to get really good at pipetting? Or using a microscope? Or planning and carrying out well-designed experiments? These are key questions we should ask when we develop a laboratory course.

Most introductory biology laboratory experiences appear to be brief surveys of a smattering of techniques, from enzyme assays, to microscopy to gel electrophoresis to taxonomic recognition and identification. At the same time, the laboratory experiences are designed to try to reinforce key biological concepts such as membrane transport or cell structure or the consequences of mutations. To achieve both disparate goals, most laboratory instructions include guiding questions that are supposed to probe the understanding of concepts and, in the same document, provide step-by-step instructions on how to use the equipment or how to do the chemical extraction in order to make the measurements for that particular exercise.

From most students' perspectives, the laboratory that accompanies the introductory biology course is interesting but not compelling. Many students feel that the laboratory is the easy part of the course, requiring only a quick read of instructions just before the session begins and putting together a chronological laboratory report or turning in a worksheet at the end. While students for the most part like working with their hands, seeing specimens first hand, they are not "doing" science. They are filling out a worksheet or data table, working toward the assignment, putting in the time. When students have the time and opportunity to conduct experiments of their own design, particularly when they have the opportunity to repeat the experiment to gain practice and expertise, they are much more invested in the techniques and also appear to look beyond the technique to the science they are doing. The goal of this book is to provide strategies, examples, and ideas for how to reorganize your introductory biology or life sciences course to give students those discovery opportunities.

In order to provide students with opportunities for both self-discovery and real scientific discovery, the laboratory sequences should be grouped into conceptual units. The emphasis must be on the discovery process, rather than on learning a skill or technique to achieve an accurate, predetermined result. While skills are important, they are the means to the end. Scientists learn new skills and techniques not to add to their list of accomplishments, but to enable themselves to ask and answer conceptual questions. To provide our students with genuine scientific training, we need to restructure how we teach these skills and techniques to better mimic the actual scientific process.

In the first few chapters of this book, I describe ways to restructure introductory and intermediate-level life sciences courses within the existing formats of most college and university introductory science courses. All that is needed is to rearrange the laboratory exercises and provide a conceptual context and framework – one that can be reinforced in the classroom part of the course. We will put the emphasis on the process of science, rather than on the content.

Then, in later chapters of this book, I suggest approaches to more fully reconfigure these introductory courses to better address the five major concepts of biology: evolution, structure/function relationships, energy transformations, information storage and transfer, and systems. Introductory biology courses need to be the place where students learn about these major concepts in order that they can be informed global citizens and leaders. Students need to understand how science is done, how scientists approach questions, and how experimental findings can further our over-all understanding of our world and our place in that world. These goals are crucial for all college students, not just those who plan to embark on careers as life scientists. The last chapters of this book explore ways to alter life science curricula both to train future leaders and educators in basic science literacy and to train future scientists. A reinvigorated introductory curriculum will not only better prepare future life scientists and medical professionals, but also inspire and encourage the nonscientist students to delve more deeply into science and will thus enhance science literacy even further.

Further Reading

1. Vision and Change in Undergraduate Biology Education: A Call to Action. 2011. American Association for the Advancement of Science. ISBN: 978-0-87168-8. http://www.visionandchange.org/
2. CBE: Life Sciences Education. 2014. American Society for Cell Biology. Online ISSN: 1931-7913.
3. Brewer, C.A. and Smith, D. (Eds.) Vision and Change: An Undergraduate Biology Education (2011) American Association for the Advancement of Science.
4. Vodopich, D.S. and Moore, R. (2014) Biology Laboratory Manual, 10th Edition; McGraw-Hill Companies, Inc. New York, NY.

2 Changing Goals and Outcomes in Introductory Life Science Course Laboratories

The Introductory Science Course Experience That We Have

Lecture sessions two to three times a week for about an hour. Laboratory sessions once a week for 2–4 hours. Textbooks that are encyclopedias of well over a thousand pages, organized from molecule to ecosystem (Table 2.1). Examinations stressing the importance of the right answer, of the definition over conceptual understanding. Classrooms crowded with students sitting in rows. Laboratories designed with long rows of standing height benches. Students following detailed procedures and filling out worksheets or turning in laboratory reports that follow specific guidelines about format over content.

How Science is Actually Done

Tables and conference rooms where scientists gather to share information, to talk about experimental results published by themselves and others, and to plan new experiments to address an overall conceptual question arising from those discussions. Weekly meetings to read, discuss, and present scientific papers, workshops to learn new techniques, and time to practice and try out experiments, revise, and repeat. Computers for data analysis, to look up facts and previously published work. Laboratory mates brainstorm new ideas and discover how each others' projects inform the others.

The individuals involved in the scientific work find it relevant and have a vested interest in seeing it succeed. Graduate students learn about a new field of study and acquire a depth of knowledge by reading experimental articles, by talking with colleagues, and by applying their knowledge at the laboratory bench. Textbooks are used only as reference sources to help

Discovery-Based Learning in the Life Sciences, First Edition. Kathleen M. Susman.
© 2015 John Wiley & Sons, Inc. Published 2015 by John Wiley & Sons, Inc.

Table 2.1 Current Structural Organization of Most Introductory Biology Textbooks

	Overall Topic	Subtopics
Unit 1	Introduction to Biology, the Scientific Process	Overview of what biology is, how to do science
Unit 2	Chemistry of Life	Atoms, molecules, bonds, water, pH, macromolecules, enzymes
Unit 3	Cell Structure and Function	Procaryotes, eucaryotes, membrane transport, cell signaling, cell division
Unit 4	Genetics, Gene Expression, Genomes	Chromosomes, DNA replication, gene expression, Mendelian genetics, genomes, DNA technology
Unit 5	Evolution	Selection, drift, Hardy-Weinberg, species, geologic and evolutionary time, phylogenetics
Unit 6	Diversity and Groups of Organism	Tree of life, major organism groups
Unit 7	Plant Structure and Function	Tissues, life cycles, major groups
Unit 8	Animal Structure and Function	Tissues, life cycles, major groups
Unit 9	Ecology	Succession, conservation, human impacts

understand the articles, much the way an encyclopedia or a dictionary might be used in other fields. Learning and understanding come in layers, with scientists-in-training becoming increasingly sophisticated in a field over time, through practice, repetition, and immersion.

Can we use actual science training as a model for developing literacy in biology in the first place? Can we dispense with the textbook-based, lecture-heavy, definition-rich focus that's been entrenched for decades?

Science educators need to change the format and the foundational experience at the introductory and intermediate levels to make learning about life science more relevant, more interactive, more like the actual experience of doing science. The study of life needs to be relevant to each and every student. We need to move away from a content overloaded, technique-learning focus. But enacting such large-scale change is difficult. The scientists who are teaching the courses have been trained to teach using the old approaches, so trying something new without a framework to follow is intimidating. In addition, most science professors have little incentive to make drastic changes. Change is time-consuming. Textbooks will take some years to change to fit new ways of teaching and learning, especially if those new ways are not broadly implemented. Academic course schedules are not easily changed because of institutional inertia.

Why do we need to change the way we teach biology? Not because we want to churn out zillions of new scientists. But, rather, because we need

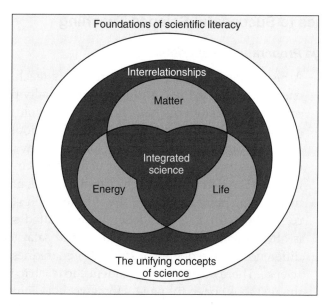

Figure 2.1   Goals for science literacy. (Image reproduced with permission from http://ae.gov.sk.ca/evergreen/science/part3/portion04.shtml.) (See insert for color representation of this figure.)

our future leaders and citizens to understand the fundamentals of science and biology in order to make informed decisions, in order to appreciate our place on this planet. We have many challenges facing us globally, from the impacts of climate change on global economies and the food supply, to combating emerging diseases, to appreciating the global relationships between us and the other organisms around us. We *need* a scientifically literate populace (Figure 2.1).

It's a super tall order. But one step toward that lofty goal is for college science educators to teach courses that engage students rather than turning them off to science. Teach them in a way that transforms how they approach science and their everyday lives. All college graduates need a solid grounding in life science and the scientific process, whether they go on to become lawyers, business leaders, government workers, or teachers. Critically, if our future teachers have an understanding and appreciation for science, they will in turn teach science from that perspective, which will in turn improve the learning and science education of subsequent generations.

There are a number of challenges our introductory courses need to grapple with. We need to understand these challenges in order to come up with effective teaching strategies.

Challenges to Successful Science Teaching

Pre-College Preparation Disparities

A big challenge facing college professors teaching introductory life science courses is that not all students are equally prepared for college-level science. Some students have not had any biology instruction since general science in middle school. Some students took AP Biology 2 or 3 years ago, and some just completed AP Biology a few months earlier. Some students have never had formal instruction about evolution because of reluctance on the part of their high-school teachers or schools. The teachers in many cases have had little formal training in science. These disparities, combined with state and federally mandated standardized testing that is often linked to teacher evaluation and salary, have created enormous differences in the degree of student preparedness to engage college-level science. There's a disturbing tendency to teach to the test or just go through the textbook, page by page. Homework assignments involve filling out worksheets where students copy verbatim from the textbook. In one ear and out the other is what happens to many students. They memorize to take the test and then promptly forget 90% or more of what they supposedly learned.

It is definitely a challenge, then, to teach introductory biology or any introductory science course, even without considering large-scale changes to accommodate the new biology. I believe that some of the new approaches described in the following section will address not only the new biology, but also the bimodal distribution of student preparedness.

Avoiding the Textbook as the Organizer of Your Course

Most of us began our scientific learning with textbooks as the font of knowledge. Our professors presented the material from the textbook by lecturing. The good ones reorganized the material and explained difficult concepts clearly, drawing on the board with multicolored chalk and peppering their lectures with personal stories and anecdotes. Most introductory science courses still use textbooks as the primary reading material, having a chapter or parts of chapters assigned to be read before each class session. Many of us can still remember trying to read 30–50 pages of complex conceptual material filled with jargon and technical terms. Do you remember getting out your highlighter and highlighting every other sentence? Or taking notes while you read, realizing halfway through your assignment that you simply do not have time to finish reading? Or, perhaps, if you had a college experience similar to mine, you began falling asleep by the third page of the assigned reading and spent your time starting over from the beginning several times before just giving up on it, skimming the headings to be able to wing it in class, deciding to *really* read it before the examination, but then never following through. Nowadays, textbooks are huge encyclopedic tomes numbering over a thousand pages, at least twice the size of textbooks 20 years ago. The books are filled with "inquiry" questions, colored boxes with historical experimental stories and key discoveries, multicolored pictures and schematic diagrams, and lists of further reading. Many books also come with online materials such as videos, animations, and virtual experiments. Although wonderful resources, there is simply too much material – information and multimedia overload. For many students, as the semester progresses and the workload increases, the textbook reading falls off. Students find themselves skimming very quickly through the assigned reading to get a quick idea of what will be covered in class, saying to themselves that they'll catch up on the reading before the next examination. Then, when it comes to the weekend before the examination, the already sleep-deprived students start studying by reading the chapters in the textbook, never really finishing all the pages assigned. Never doing all the ancillary material designed for active engagement.

While having an introductory textbook associated with your course is an essential resource for your students, its place is better as a reference, rather than as the scaffold for your syllabus. Students need to learn how to use their textbook. Rather than sitting down to read the chapters like they are novels, paying little attention to the figures, diagrams, boxed information, and guiding questions, those ancillary features on each page should probably be the focus of their reading. And certainly, for instructors of introductory courses, we need to break free of the textbook as the starting place for our syllabuses.

If you take the textbook out of your course structural framework, you will probably find it easier to identify integrative concepts that link fundamental concepts together at multiple levels of biological organization. Your introductory course will then be much more compelling, organized around one or two exciting and relevant topics. For example, a topic that works very well for teaching lots of introductory biology concepts is cancer. Cancer is a cellular process and so lends itself to discussions of topics such as cell communication, DNA replication, and mutations. In addition, cancer is a tissue/organ and organismal disease. You can cover things such as various tissues and physiological function of organs and organ systems. Cancer is also a population phenomenon in the sense of selection pressures operating on tumors and on cancerous cells during chemotherapy. So you can teach topics such as natural (or artificial) selection and evolutionary change. If you couple your cancer topic with a second topic such as "biological impacts of climate change," you could also consider topics in ecology, ecosystems, biogeochemical cycles, and the like. I recommend two topics as organizational themes for an introductory course. If you take a topical approach to teaching fundamentals of biology, then your readings could be taken from the secondary literature, including review articles or more general science articles, rather than a 1000-page textbook. The textbook could serve as your encyclopedic reference for terms and concept definitions. Many textbooks also describe the early and key experiments to provide an historical context that enriches and deepens the learning experience.

Weaning Away from Content-Heavy Lectures

Think back to how you learned your current field of specialization. For me, the real learning and mastery began with my senior independent research project, not with my courses. Sure, I memorized a bunch of stuff in my introductory- and intermediate-level courses. I even did some pretty interesting laboratory exercises in a few of my upper division courses. But nothing really stuck for longer than a few months. My view of the world around me wasn't transformed, nor was I. I felt like I had acquired a tiny smattering of factoids but little in the way of deep understanding. Then, senior year, everything changed.

I undertook a senior research thesis. Because it was up to me to figure out how to do things, I approached the material differently. I read the experimental and review articles in the field because I knew that I needed to be able to apply that information in designing my own experiments. The articles were incredibly dense and difficult for me to understand. I read those articles over and over, looked terms and concepts up using the index of my textbooks and my old class notes, and slowly started to understand. I taught myself the concepts. I designed an experiment, using those research articles as a guide, to investigate some tiny aspect of a question I was curious about. My professorial mentors showed me how to perform a few key techniques, and I paid really close attention because I was planning to use those techniques on my own to address my own experimental question. Then, over the course of weeks and even several months, I performed experiments using those same techniques, over and over, refining, revising, and re-doing. I became much more interested and invested in the course material I was learning in my other biology courses. *This* is when I learned how to do science; when I learned some of the key concepts in biology. My last year in college. It shouldn't be that way.

Obviously, we can't offer these kinds of experiences to all our students. But if we analyze the elements that made the learning happen, I believe we can incorporate those elements into our classroom and teaching laboratories.

The Elements of Successful Science Learning

Student Autonomy

When students are empowered to develop their own ideas, to explore their own interests, they read and think about the course material much differently. This is particularly the case if the students have to apply what they are learning to a project they feel invested in or to solve problems that are relevant to their lives. The laboratory portion of your course can foster self-discovery and autonomy by incorporating independent small group projects. If students are invested in an independently designed series of experiments or structured explorations, they are more motivated to understand the concepts operating. Terms, facts, and concepts related to their laboratory explorations are learned more completely and more naturally. The laboratory experience is the first step toward constructing the conceptual scaffold upon which the students will build their knowledge and expertise.

Relevance

Why do students enroll in introductory biology courses? Mostly because it is required, perhaps to fulfill college general education or distributional requirements. Because they want to go to medical school and become doctors. Because they want to get involved in responding to climate change. Because they love hiking or animals or plants. Because they like the idea of science. Sometimes, they come to biology curious and hoping to learn more about their own areas of interest. Unfortunately, so often, these varied interests die on the vine in the face of content-heavy lectures, lists and lists of vocabulary words, requirements to memorize pathways, and concepts that do not appear immediately relevant to their original reasons for taking the course.

Imagine a course that is tailored to student interests or one that sparks new interests. Imagine such a course at the entry level. Sounds pretty exciting, doesn't it?

Student Investment

As a kid, I had always loved watching *Jacques Cousteau* specials on TV. I was thrilled to see the incredible fish, corals, swimming dolphins, and scary sharks. I would daydream about learning to scuba dive and lounging about in a catamaran after a fulfilling dive in the Caribbean discovering new species or new interactions. When I went to college, I wanted to become a marine biologist. Later on, as an adult (with a very different career, but still a biologist), I finally decided to learn to scuba dive. I paid attention to lectures about oxygen under pressure, about how to operate pressurized air tanks, and about the frightening physiological mishaps that result from not understanding the underlying chemistry, physics, and biology of scuba diving. I learned quickly and permanently. Later, when I learned CPR, I retained far more of the biology and physiology than I ever had when sitting in my physiology lecture learning about the heart and lungs and struggling to stay awake during long, fact-filled lectures. What's the difference?

One big difference was my level of investment. I cared about learning the material the *first* time. I was there, learning *now* to apply the concepts first-hand. I was taking notes, but my goal was not to take notes now and learn it later. The goal was to learn it *now*. My whole attitude toward learning was different than when I was sitting passively in lecture hall, taking notes as the material flowed by me. We need to bring that level of investment to our classrooms. We can do that by making the material relevant to the students, to having them feel invested in being there in that class at that moment. That they need to get it now, not in 3 weeks or 3 years later for the MCAT.

Sustained Engagement

We all know that we retain what we've learned better if we work with it, interact with it, for a sustained period of time. *Listen, read, plan, do, reflect, revise, and repeat.* A key aspect of this learning sequence is the repetition. Most introductory biology laboratories do not offer the opportunity for repetition. One week is spent learning how to perform a spectrophotometric assay. Or, student groups conduct one limited experiment examining the effects of changing pH or temperature on enzyme function, spending perhaps 2 weeks on the exercise. One run-through of the experiment and then time to move to the next technique. A true mastery of the technique, a firm understanding of the biology happening during their "experiment," cannot be achieved with this one-pass approach. In allowing students to repeat an experiment or to design a second experiment using the same techniques, we give students the repetition and a more sustained engagement with a technique or experimental approach necessary for long-term understanding. Adding in repetition means you will have to give up some material or some techniques in your introductory laboratory. You have to be choosy, plan wisely. So, the first step is to decide how you want to organize your overall course.

Understanding Through Teaching

What better way to truly master a concept than to teach it to someone else? To teach someone, you have to be able to think about it from different perspectives, to break it down into manageable chunks that you re-organize and manipulate. You need to be able to put it in your own words. If you can articulate what something means to you, then you can teach it to someone else. The actions involved in teaching include a mental attention and manipulation of a concept in your head and the organization of this material into spoken (or written) form. It's not just recognition of terms in a book. It's an active, effortful experience. When you teach someone, your goal is to have your pupil understand. This means that you are looking for feedback from your pupil. In response to this feedback, you revise what you say, and you try different approaches. These actions are all revisions, repetitions, and practice, which cement your own understanding even as you teach someone else.

The more of these elements that you incorporate into your introductory course, the more successful the learning will be. Discovery-based laboratories emphasize all of these elements. What follows are two different organizational strategies for an introductory biology course that has a laboratory component. Then, Chapter 3 provides a more in-depth example of an actual introductory biology laboratory course currently in use.

Two Re-organizational Schemes for an Introductory Biology Course

Re-organizational Scheme 1: Putting the Classroom First

The majority of us are most comfortable, when we plan our introductory science course, to think first about the classroom. What concepts do we want to cover? What content do we want our students to learn? What order do we want to present topics? The answers to these questions result in our course plan, our syllabus.

A lot of us might begin to answer these questions by leafing through a few introductory textbooks. As we feel our students need a textbook (because there's so much to learn, and our students expect this), we are often unwittingly influenced by the textbook's organization of the content that we hope to teach. After all, we were taught that way, and we are successful scientist-educators! But, of course, the vast majority of our peers who were educated this way did not choose to become biologists or educators. Many were turned off to science by their introductory experiences.

Let's try to break free of this pattern that is so comfortable. We'll start slowly, just easing ourselves into these reorganizations. Firstly, let's start with the five major conceptual areas I noted in Chapter 1. I've rearranged them a bit, thinking about how I might teach a class around these themes without moving too far away from our comfort zone. The order of topics in this case is basically what we are all familiar with from the major biology textbooks on the market, just recast a bit. A key difference, as we'll see, is that each theme or concept will reflect multiple levels of biological organization, rather than the march up levels that is common in the textbooks.

(1) Structure and Function
(2) Information flow, exchange, and storage
(3) Pathways and transformations of energy and matter
(4) Evolution
(5) Systems.

Now, let's take a look at how the laboratory portion might provide support for these conceptual topics and goals. Table 2.2 groups laboratory sessions around those conceptual themes.

In this scheme, the laboratories provide practice with the concepts and juxtapose the levels of organization to illustrate interactions; however, they are not really discovery-based. Rather, the laboratory experiences are inquiry-driven and support the classroom learning. The examples of laboratory exercises shown in Table 2.2 were taken from published introductory biology laboratory manuals (e.g., Vodopich and Moore, 8th Edition; Dolphin, 8th Edition; see Further Reading). What I'm illustrating in this case is that you can take laboratory sessions that you have used in the past or that are published for use in large-scale university course settings, and you can rearrange the order, group them, and adapt them to a more modular format that can support a conceptual approach in the classroom. The laboratory sessions will then naturally connect through the overarching concept. For example, you might have a 2-week module of structure–function relationships. In the first week, you might have students learn to prepare microscope slides, to observe plant cells, and also to examine plant pigments found in chloroplasts. Students could design short discovery-based experiences that examine chloroplast motility or cytoplasmic streaming in response to changes in temperature or light intensity or exposure of the plant tissues to herbicide. Or they could evaluate the effects of ion concentration or herbicide on plant cell wall

Table 2.2   Putting the Classroom First: Conceptual Organization

Conceptual Theme	Skills and Competencies	Levels of Biological Organization	Grouping of Existing Laboratory Exercises (Based on Published Manuals)
Module 1: The relationships between structure and function	Microscopy, comparative anatomy, field observations	Cell, tissue, organ, organism	Lab 1: cells and tissues structure/function laboratory (How do cells work? How do cells form tissues?) 2: Plant biodiversity of form and function 3: Animal biodiversity of form and function 4: Community structures and functions (field-based observations)
Module 2: What is Biological Information? How is it stored and transferred?	Microscopy of mitosis, meiosis; DNA/Protein electrophoresis; Mutation analysis; Mendelian analysis; sea urchin fertilization	Cell, biochemistry/molecular, organism, population	1: Comparisons of mitosis and Meiosis 2: DNA extraction and electrophoresis lab-Information Molecule 3: Human Mendelian traits (blood typing, etc)
Module 3: How do living organisms obtain and use energy?	Enzyme assay; photosynthesis measurement; Respiration; quantitative reasoning	Biochemistry; molecular; cell; organism; interactions with primary producers	1: Either respiration (eg. Yeast fermentation) or photosynthesis lab (absorption spectra of photopigments; starch production in leaves, etc) 2: Enzyme assay using spectrophotometry-independent experiments 3: Nutrient cycling lab-measurements of soil, water
Module 4: Evolution and Adaptation	Bacterial growth; microbial techniques; fruit fly traits; populations study; quantitative reasoning; Mendelian genetics	Organism, population	1: Mendelian and Hardy-Weinberg population modeling lab 2: Antibiotic resistance bacterial lab 3: Determining genotypes in fruit flies
Module 5: The Interactions and Interconnections of Living Systems	Field-based studies of biodiversity; simulation; modeling	Population; individual; ecosystem	1: Field survey of plant community succession 2: Bacterial growth limitations or duckweed growth measurement lab 3: Brine shrimp growth experiments

integrity. Or, they could compare plant cell structure/chloroplast density from plants grown in different environmental conditions or plants of different species. Because this laboratory module would occur early in the term, you will probably want to provide the students with specific ideas or choices of questions for them to explore. Have them design their exploration – choose the temperatures or the light levels or the times/concentration of exposure. This way, the students begin to practice how to design an experiment and how to take a question of interest and use a scientific process for examining that question. The following week, students could explore structure–function relationships at the level of whole plants or plant communities (Table 2.2). If you have the students read a primary experimental paper in conjunction with the laboratory or linked to topics addressed in the classroom, you will have approached the key concept of structure–function relationships at more than one level of biological organization, in both the classroom and the laboratory. The laboratory's emphasis on student self-discovery will much better reinforce these conceptual goals.

The next conceptual area might be energy transformation. To support this topic in the laboratory, students could examine photosynthesis and respiration. You could take this overall organization a little further by grouping two or more of the conceptual areas into larger, several-week long experiences. For example, you could group Structure–Function with Energy Transformations into a 5- or 6-week long laboratory module. You could add in a couple of weeks of an enzyme assay or photosynthesis/primary productivity assay that students could design experiments to discover. Another example would be to combine Information Storage/Transfer with Evolution, as presented more fully in Chapter 3.

Re-organizational Scheme 2: Putting the Laboratory First

What if the classroom was there to support the laboratory? This is a markedly different way to organize an introductory biology course. At my institution, our required introductory biology course is in fact a laboratory course that centers on two or three large topics. The classroom portion, which is greatly reduced compared to the 3 hours per week in class at most institutions, supports the laboratory by providing background information or opportunities for data interpretation or analysis. Chapter 5 presents this model for introductory biology instruction in much more detail. This section sketches out some general possibilities for how one might structure a more standard biology course to center on the laboratory rather than on the classroom experience.

What would an introductory biology course look like if the focus was on the laboratory rather than on the classroom? This organizational scaffold would most definitely break us free of the content-driven curricula that we're trying to get away from. The course would become more modular, more topical. The key is the choice of topic.

Topics would need to support student-driven discovery-based experimental approaches that are feasible and accessible to introductory students. Topics would also need to support multiple levels of biological organization and should at least introduce the major core conceptual areas. This is a tall order for an introductory laboratory-focused course. But doable.

Chapter 5 articulates a course that we have had in place for over a decade. It's fairly successful as an introduction to the study of biology and the life sciences. We certainly haven't seen a decrease in the number of students interested in taking our courses, quite the opposite in fact. In this chapter, I briefly outline some possible topics and laboratory sequences and then show how the classroom portion can support those sequences.

Example Topic: Biological Arms Races (Conceptual Areas: Structure and Function, Information Storage and Transfer, Evolution, Systems)

A farmer plants a crop of corn, only to have it devoured by insects. So, the next season, the farmer plants a crop of corn seed that has been coated with a pesticide that will be present systemically within the plant and so will repel those insects. If the farmer plants the same corn seed year after year, the insects develop resistance to the pesticide. The farmer then plants a different seed stock that is coated with a different pesticide. Or, the farmer combines the coated seeds with a liberal application of pesticide sprayed from a plane.

A patient is diagnosed with a bacterial infection and is prescribed an antibiotic. However, the patient only takes a couple of the pills and stops once he feels a little better. After a few weeks, the infection returns, and the patient reaches in the medicine cabinet and takes a few more of the original batch of pills. This time, the pills don't work as well, so the patient returns to the doctor and gets a new, slightly stronger prescription and does the same thing. The patient, still contagious, happened to infect several others, whose infections appeared resistant to the antibiotic in the dose prescribed for the original infection. Some of those patients take their medicine for its full course, but a few do not, forgetting to take their pills once they feel a bit better. Eventually, the bacterial infection is completely resistant to that antibiotic and a different one needs to be prescribed.

What Do These Scenarios Have in Common? What is Going On?

In the laboratory, students perform an antibiotic resistance project. They learn in the classroom about how antibiotics work, why they work on prokaryotic organisms but not eukaryotic organisms, what antibiotic resistance is, and how it arises. They learn about microbial diversity by considering how different antibiotics affect some groups but not others. They learn about gene expression, regulation, and mutation. They can learn about population-level microevolutionary change and can model population-level changes in gene frequency. The classroom can link what they are doing in the laboratory with bacteria to other systems such as pesticide- and herbicide-resistant organisms, toxin resistance, and natural biological arms races between predators and their prey. The laboratory can evaluate how rapidly these changes can occur and can serve as a model for how similar co-evolutionary processes occur in nature.

The laboratory sequence might unfold as in Table 2.3.

Table 2.3   Laboratory-organized Class Schedule

Week	Topic	Activities	Core concept Addressed/ Competency
1	Exploring the microbial world		Microscopy
2	Measuring growth		Density measurement, sterile technique, colony exploration
3	Exploring antibiotic resistance - how do microbes resist antibiotics?	Antibiotic resistance assay; experimental design	Experimental design, colony exploration, measurement/dilution Classroom: bacterial resistance strategies
4	Exploring antibiotic resistance	Student-designed experiments	
5	Continuation	Experiments repeated or extended	
6	Modeling evolutionary change	Data analysis, application of a mathematical model	
7	Structure-Function: mutations	PCR analysis of antibiotic resistance genes	DNA extraction, PCR

Classroom Support for the Laboratory Work

Week 1: Exploring the microbial world. What kinds of organisms constitute the microbial world? In the classroom, you can compare prokaryotic and eukaryotic organisms. A central question might be the targets of antibiotics and why they work in prokaryotes, but not in eukaryotes. The classroom can also support the laboratory by addressing the cell structure and function comparisons of prokaryotes and eukaryotes, focusing on the different types of mechanisms that antibiotics function. For example, the ways that antibiotics cross bacterial outer membranes (via porin channels). Another mechanism by which bacteria evade antibiotics is via production of cellular transport pumps, such as the multidrug resistance pumps. Concepts such as membrane transport and membrane structure/function can be addressed within this topic. Students can explore the diversity of prokaryotic organisms within the context of the mechanisms by which different antibiotics operate, as well as within the context of how bacteria evolve resistance to these antibiotics.

Week 2: Measuring microbial growth. This topic can be supported in the classroom by considering the dynamics of population growth and the mechanisms of cell division (mitosis, the cell cycle, and meiosis). This topic could also encourage a consideration of DNA replication, transcription, protein synthesis and mutations, genes, and alleles. Bacterial growth can be compared to the growth rates of eukaryotic organisms, both unicellular and multicellular. Population dynamics and competition for resources are concepts that fit in here as well. Rates of evolutionary change depend on population size and population growth rates, and so these issues can be discussed.

Week 3: Antibiotic resistance. Continuing this topic within the classroom, you can continue to explore different mechanisms of antibiotics. Many enzymatic pathways exist to detoxify antibiotics. A discussion of some of these mechanisms can enable you to address core concepts such as energy/matter transformation and enzymatic pathways. You can address how rapidly these detoxification pathways can arise in bacterial populations via horizontal gene transfer and other mechanisms. The speed with which resistance develops within a population of bacteria depends on the population size and the genetic variation within that population, as well as the "strength" of the selection pressure. Thus, within this discussion, you can consider core concepts such as natural selection, population-level change in gene frequency (e.g., Hardy-Weinberg), and mechanisms of gene flow in prokaryotes versus eukaryotes.

Week 4: Antibiotic resistance experiments. You can continue to support the laboratory by discussing ways in which gene mutations can lead to antibiotic resistance. A number of resistance mechanisms involve the chemical modification of the antibiotic target, which lowers the affinity of antibiotic binding or alters the production of essential precursor compounds by altering transcription or translation activity. These different mechanisms will allow you to consider structure–function relationships at the level of molecular structure, gene expression, cell growth, and population dynamics. You can also address the relationship between antibiotic resistance at a cellular level and at a global clinical level. How does the evolution of antibiotic resistance progress from the level of individual mutations that enhance an individual's fitness to a global trait of a species or group of bacteria? You can expand the topic here to compare with other types of evolutionary change, such as artificial selection of pesticide resistance, or artificial selection used in developing agricultural varieties, dog breeds, and the like. You could also consider examples and mechanisms of evolutionary change such as genetic drift and gene transfer. These issues can be integrated beyond prokaryotes to consider the evolution of other characteristics and traits of other organisms.

Week 5: Antibiotic resistance experiments, continued. To link the laboratory to the classroom while students are deeply involved with their independent projects, you can have students read a current experimental paper on an aspect of antibiotic resistance or another example of evolutionary change. At this point in the semester, students should be able to integrate what they have learned at multiple levels of biological organization and should have a good handle on evolutionary change. When students are exploring the differences between prokaryotic and eukaryotic cells in the laboratory, you can have your students read about how bacterial infections arise and the different types of immune defense employed by multicellular organisms. You can examine structure–function relationships at multiple levels of biological organization. When the laboratory introduces students to topics of growth and metabolism, you can consider these topics in the classroom, discussing mitosis, meiosis, and gene transfer, as well as mechanisms of cellular metabolism shared between prokaryotes and eukaryotes. You can also introduce the idea of antibiotics and explore the mechanisms of action of common antibiotics, many of which exploit the differences between prokaryotic and eukaryotic organisms. Each of the five conceptual areas can be explored within this topic of antibiotic resistance/biological arms race.

Summary

This chapter shows how you can rearrange your standard introductory biology course into a more cohesive, topics-driven experience. There are many different ways to organize these topics, and there are many differ-ent topics that would engage your students while they focus on antibiotic resistance in the laboratory, so you can tailor the classroom support to your own areas of interest and expertise and overlay those interests over the five overarching concepts: Evolution, Energy Transformation, Structure and Function, Information Storage and Transfer, and Systems. You can have your classroom drive the organization of the laboratory, or the other way around. Rearranging laboratory skills into multilevel groupings organized under the umbrella of your chosen topic or under the conceptual areas will switch the focus from skills and techniques to active inquiry and discovery. The following chapter demonstrates how to implement these laboratory ideas.

Further Reading

1. http://amrls.cvm.msu.edu/microbiology/bacterial-resistance-strategies/introduction
2. Vodopich, D.S. and Moore, R., (2014) Biology Laboratory Manual, 10th Edition; McGraw-Hill Companies, Inc. New York, NY.

3 Incorporating Discovery-Based Laboratory Experiences at the Introductory Level

How do you develop a series of laboratory experiences for your introductory biology course that complements your pedagogical goals? The first step is to clearly articulate what your goals are for your introductory biology course. Do this first for yourself, so you have a clear idea of where you want to be headed with your course.

Most introductory biology courses introduce core concepts in biology, including evolution, information flow, energy transformation pathways and cycles, cell structure, function, and communication, and organism and community structure, function, and interaction. Because these courses are the foundation for a curriculum that seeks to graduate citizens with some depth of knowledge about biology, the hope is that students will master these concepts and principles and deepen their skills and content knowledge in subsequent courses at the intermediate and advanced levels.

At the same time, most introductory courses also broadly survey immense fields of knowledge. Every faculty member who teaches an intermediate- or advanced-level course in a subfield of the discipline expects some introduction to that subdiscipline and its fundamental concepts at the introductory level. What results is a survey course that has breadth without a lot of depth. Most introductory biology textbooks take students through a taxonomically organized survey of life's diversity, from bacteria to protists to fungi to plants to animals, but this march through life is not even on the list of major goals for most introductory courses. It sure appears to be a crucial aspect of introductory biology, from looking at the available college-level textbooks out there.

In addition to the core concepts, introductory biology courses with laboratories stress the importance of quantitative reasoning and give students exposure to the rudimentary skills of biological investigation, including measurement, graphing, microscopy, and, increasingly, experimental design. More recently, science educators believe that students need to

Discovery-Based Learning in the Life Sciences, First Edition. Kathleen M. Susman.
© 2015 John Wiley & Sons, Inc. Published 2015 by John Wiley & Sons, Inc.

learn how to apply the scientific method by undertaking "authentic" research experiences and by communicating those experiences in different ways. How do you teach a broad array of fundamental laboratory skills while allowing for discovery or inquiry?

Can introductory courses effectively satisfy all these goals? In general, educators and students alike are dissatisfied with the past half century of status quo. The existing structure has turned off generations of college students to science and implies that doing science is a matter of following procedures that resemble recipes in a cookbook. The dissatisfaction of both educators and students combined with the explosion of new data and new approaches in biology, as well as increased public awareness of the need to understand the biosphere and how our activities are affecting other organisms and systems, has led to the articulation of a new set of curricular goals for college-level biology.

These goals do not stress breadth or survey available knowledge. Rather, they stress a way of approaching knowledge. The goals build a scaffold, a framework, on which to position knowledge about organisms and systems. These curricular concept goals should guide you as you develop your introductory course. These goals are appropriate for introductory, intermediate, and advanced courses in the life sciences. The key in this case is articulating goals for an introductory level. An example might be illustrated by the following questions posed in an introductory biology syllabus:

- How are biological systems organized?
- What is the relationship between structure and function in biological systems?
- What is biological information? How is it stored and exchanged?
- How do biological systems acquire and transform energy?
- How does biological diversity arise?

The questions construct a scaffold for subsequent, more advanced work in biology. We learn best by integrating our previous knowledge onto an existing framework. So, if we set up the scaffold right at the beginning, students should be able to add to their knowledge and to deepen their understanding by accessing their foundational conceptual framework. Attempting to survey a vast array of facts using an intellectual framework that is not yet built will not lead to long-lasting learning or understanding of the underlying concepts. You need to be careful of what examples you choose and which content to provide and elaborate upon, especially at the introductory level. This approach runs counter to how most introductory biology textbooks are organized.

Let's take an example of how one might proceed: the concept of biological information and how it's stored and exchanged.

Current introductory biology textbooks have multiple chapters devoted to DNA and how genes are expressed and regulated. There are multiple chapters on how DNA is exchanged in bacteria and eukaryotes through fission, mitosis, and meiosis. Multicolored diagrams, examples of key experimental discoveries, accompanying videos, and lists of further reading make for such a busy reading experience that students are soon overwhelmed by details such as Okasaki fragments, topoisomerase, origins of replication, TATA boxes, and mitotic spindle proteins. Glorious, wonderful details, but the vastness of the details for a first-year student leads to information overload. Do they really need to memorize all these terms? What's really important to remember, to understand? Can you find the forest for the trees? What does an introductory student, or one for whom this is their only science course, need to know? Where is the framework, the scaffold? All too often, biology professors are afraid that leaving out details will dumb down the biology. Sometimes, the students who took AP Biology in high school pressure their college instructors to provide more detail than they got in high school, especially if it's the same textbook. If the textbook provides so much detailed content and the instructor doesn't delve into it, some students feel short-changed or wonder when they will actually *learn* all that detail. If it's there in an introductory textbook, doesn't that mean that it's essential to learn in order to proceed to more advanced learning in biology? I don't think so.

Maybe an analogy would work in this case. How do we best learn to play a musical instrument such as the piano? You start off with simple, familiar tunes, where the "notes" on a piece of music paper might be numbers that represent your fingers. First one note at a time. Simple, familiar melodies where you focus on how to position your hands and fingers. Then, you begin to learn the components of simple chords (the first, third, and fifth) and the notes of a scale. You don't start out with a full Beethoven sonata (Figure 3.1)! Why do introductory textbooks provide so much detail? If you read the beginning material in most introductory textbooks, the part where they tell you the mission of the book, you will often read that one purpose of the text is to serve as a reference later on. To justify the very high price of the book, the authors and publisher reassure us that the book will still be a useful resource when students are in graduate school! This means that the introductory textbook is really not an introduction to the discipline; it's an encyclopedia.

If, as a beginning piano student, you were given the music to Beethoven's Sonata No. 28 in A Major and told to just look at the top row of notes and ignore all the details, would you be able to use this

Sonata No. 28 in A Major

Theme from **Beethoven's Symphony No. 9**

Figure 3.1   Which piece of music would you want to use when beginning to play the piano? (Top: Beethoven_Werke_Breitkopf_Serie_16_No_151_Op_101.pdf. Public domain music site: http://imslp.org/wiki/Main_Page. Bottom: Courtesy: M. Grayburn. recordersupport.weebly.com/ode-to-joy.html.)

music as your introductory text to learning to play the piano? I don't think so.

Okay, so how can you design your introductory course so that your students first build a conceptual framework and then fill the frame with content? I would recommend you to start by considering the textbook to be an encyclopedia, not a reading resource like I described in Chapter 2. Do not organize your syllabus around the structure of a textbook. Do not assign your students to read particular chapters for each class session. Instead, create a course structure that builds the scaffold that your students will retain and build upon in the future, whether they are continuing on to become a research scientist or becoming an elementary school teacher. Assign readings, such as Scientific American articles or mini-reviews from journals such as "Current Opinion in Cell Biology" or "Trends in Environmental Science," which illustrate key concepts or issues relevant to the students and use the textbook as a language dictionary or background encyclopedia. The figures and accompanying material in most textbooks are really valuable for visualizing difficult ideas and content. Teach your students how to use the index. Chapter 5 has an example introductory syllabus that will give you an idea of how to do this.

You want your introductory course to reinforce the core concepts and build upon the scaffold as the semester or year progresses. This means that you need to incorporate the material that precedes it at each stage along the way, to integrate and synthesize. For example, if the first major concept that you explore with your students is how biological systems are organized, you will likely stress the levels of organization, from molecules to cells to tissues to organisms, and so on. How do these levels interact?

What are some similarities in how each level is organized? What are some differences? Why is it important to appreciate these levels of organization? Then, when you move on to the next core concept, that of structure and function relationships, it's important to approach this concept from the perspective of multiple levels of organization. Similarly, the core concept of energy transformations should be viewed through the lens of levels of organization, building in the interactions between and among levels of organization. Obviously, most current textbooks aren't really organized this way. The books move steadily "up" levels of organization, from molecules to systems, with very little emphasis on the interactions and fluidity between levels of organization. As a result of this kind of course and book organization, students will not themselves make the connections, and even if you do it in class, the influence of the linear organization of the book and your syllabus will overtake whatever you do in class.

To allow students to practice and explore these core concepts, develop laboratory exercises that include more than one level of biological organization. The most effective overall course design to accomplish these goals would be to develop laboratory experiences that also allow students to explore the core concepts in biology, a joining of the classroom and the laboratory components to a course. For the moment, let's try to develop a course that would have the current format of a separate classroom component that meets two or three times each week for about an hour and a laboratory component that meets once each week for 3 or 4 hours, as this is the current structure of many introductory college biology courses.

The best way to reinforce the concepts is to have hands-on experience with them, through laboratory experiences that are interesting and engaging and that allow students to think and ponder and reinforce. These experiences are akin to piano practice. Through doing, you reinforce your conceptual framework, strengthen and solidify it. You can also add to that framework; flesh it out. This chapter focuses on the laboratory sessions. A later chapter (Chapter 5) will join the laboratory with the classroom sessions.

The Reality of Introductory Biology Laboratories

The vast majority of college introductory biology laboratories is designed to provide a survey of basic laboratory techniques as described in the previous chapter. There are a number of published "Laboratory Manuals," often bundled with a textbook, with a series of stand-alone, single-week exercises that have little or no carry over from week to week. The exercises are interspersed with questions for students to answer, some of which might require some critical thinking. These are labeled "inquiry driven." The idea is that by having students answer thought questions or by asking students to design "thought experiments," the students will learn the underlying concepts better and will retain more in the long run. These newer, more interactive laboratory experiences are richer and more meaningful to students than a simple list of procedures, but the exercises themselves offer little in the way of the kind of sustained engagement that is important for a process of discovery. In addition, these exercises are designed primarily to illustrate a technique (such as how to do an enzyme reaction assay with a spectrophotometer) or a concept (such as osmosis or mitosis), rather than to actually give students practice in the scientific process. Newer, online resources are cropping up that provide simulations, video tutorials, and animated demonstrations of these concepts and techniques that might be an adequate substitute for doing these in the laboratory. Is there a big difference between pipetting solutions into cuvettes and placing them in a spectrophotometer and running a virtual spectrophotometric assay, other than the motor memory the real deal provides your hands? Do you learn more about why you would take a spectrophotometric approach to a scientific question from doing the pipetting yourself and filling out a chart? The answer depends on the learning goals.

In addition to the conceptual goals, life scientists want college students to master a number of "core competencies" or skills as follows:

- ability to apply the scientific process to a question of interest,
- ability to apply quantitative reasoning,
- ability to model or simulate a concept or process,
- ability to tap into the interdisciplinary nature of science,
- ability to communicate and collaborate with other disciplines,
- ability to understand the relationship between science and society.

(From: *Vision and Change* 2011 – see Further Reading)

Converting the Survey Approach to Biology Techniques into Discovery-Based Experiences that Emphasize Concepts

Try to group a number of skills and biological levels of organization into one multiweek laboratory experience that is organized by a theme(s). You could have two or three multi-week modules connected by one or several themes. This is not how most introductory biology laboratories are organized. For example, most introductory biology laboratories begin with a laboratory on making measurements of various sorts (weighing, liquid measures, graphing, etc.) and then follow with a laboratory devoted to how to use a microscope. Then, subsequent weeks reinforce concepts such as osmosis, diffusion, macromolecular composition of cells, enzyme reaction rates, and visualization of mitosis and/or meiosis. The laboratories then march up levels of organization (genetics, ecology, etc.) and across phylogenetic groupings (plants, animals, etc.). A number of these experiences can be re-grouped into broader themes that allow for student discovery and exploration, in addition to the introduction of techniques. We started to do this in Chapter 2.

The structure of most introductory biology laboratories causes students to compartmentalize the techniques and the knowledge they gain from answering the inquiry questions. Students do not see the connection between the skills developed in laboratory and the material covered in the classroom portion of the course. Indeed, in many cases, there really is no connection.

Table 3.1 shows the types of techniques and skills typically included in an introductory biology laboratory course. In the column to the right, I have indicated how these techniques and skills can be incorporated into a modular laboratory organization of two or three laboratory modules that allow students to practice the techniques multiple times while developing and carrying out discovery-based experiences. Several additional modules then follow in Chapter 5. On the basis of your interests and expertise, you can choose from among these. You can keep the course fresh by changing the order or choices of module that you include from year to year.

The modular approach does not allow much time for the animal and plant surveys usually found in the latter part of an introductory biology laboratory because that time is devoted instead to student-driven discovery experiences. Although it is fun and fascinating to see a variety of organisms, week after week of surveying groups of organisms contributes to the sense that biology is static, a bunch of facts and lists of things to memorize. Each of these weeks involves observations and answering questions, so nothing new technique-wise, simply more content and details. Not necessary for achieving the goals of the introductory laboratory.

You can design a laboratory course that involves both self-discovery (i.e., new information for introductory students, even if the actual information is not new) and new discoveries and also introduces students to the majority of the approaches and techniques that are listed previously. A growing number of inquiry-driven and discovery-focused laboratory modules are published in science and education journals. Although these are often tailored to fit a particular course or to a particular instructor's expertise, the frameworks are similar and can be modified to suit your needs.

Table 3.1   A Typical Introductory Biology Laboratory Game Plan

Lab Topic	Techniques/Skills	Covered using Discovery Modules below
The Process of Science	Hypothesis forming, experimental design, writing	Module 1, Module 2, Module 3
Measurements in Biology	Metric system, balances, liquid measures	Module 1, Module 2, Module 3
Techniques in microscopy (often two weeks)	Microscope use, Specimen preparation, observation	Module 1, Module 2
Acids, Bases, pH	Titration, pH meter, serial dilutions	Module 1
Macromolecules	Colorimetric assays, column/paper chromatography, electrophoresis, macromolecule extraction	Module 1, Module 2, Module 3
Determining solutes, products using spectrophotometry	Colorimetric assays with a spectrophotometer	Module 1
Diffusion, osmosis, cell membranes	Dialysis, osmometer, microscope observations	
Enzyme Function	Spectrophotometer, enzyme reaction rate, standard curve	Module 1, Module 3
Respiration	Volume displacement assays of various sorts	Module 1
Photosynthesis	Colorimetric assays, oxygen measurement. Volume displacement assays	Module 1
Mitosis	Live onion root tip microscopy	Module 2
Meiosis	Prepared slides, models, simulations	Module 2
Biotechnology: DNA isolation, bacteria transformation	DNA extraction, sterile technique, antibiotic resistance screening of transformed bacteria	Module 2, Module 3
Mendelian genetics	Blood typing, human traits	
Evolution by natural selection		Module 2
Diversity of plants- field ecology	Field trip to observe plant communities	Module 1
Community Succession	Field-trip observations or bacterial Winogradsky column, or milk souring	Module 1, Module 2
Population growth	Bacterial growth turbidity measurements, brine shrimp growth	Module 2
Survey of bacteria	Petri dish bacterial colony observations, gram stain, microscopy	Module 2
Survey of protists	Microscope observations	Module 2

Table 3.1 *(Continued)*

Lab Topic	Techniques/Skills	Covered using Discovery Modules below
Survey of Fungi	Microscope observations, mushroom dissection	
Survey of plants	Microscope observations, specimen dissection/observation	Module 1
Plant anatomy	Dissection and microscope observation	Module 1
Plant physiology	Specimen observation, dye transport, plant specimen analysis (eg. Phototropism)	
Survey of animals (often several weeks)	Microscope observations, specimen observation/dissection	
Vertebrate tissues	Microscope observations	Module 3
Human biology	Reflex measures, breathing, cardiac measures, sensory measures	
Vertebrate anatomy	Models, specimens, dissections	
Embryology	Sea urchin fertilization, microscope slide observations	

Taken loosely from "Biology Laboratory Manual, 10th Edition"; Vodopich and Moore, McGraw-Hill (2014).

Notice that a lot of current methodologies and approaches are missing from the aforementioned lab syllabus. Today's biologists have new tools in their tool kits, including bioinformatics approaches, use of model organisms and transgenic organisms, use of complex data sets, and use of mathematical modeling, to name a few. To fit in these new techniques and approaches, you have to remove some of the older approaches.

Module I: What are the Effects of Different Aspects of Climate Change or Other Anthropogenic Changes on Plant Primary Productivity?

An area of current interest in biology and relevant to everyday life, who isn't concerned about the impacts of global climate change, particularly introductory biology students? We all hear the news reports and claims of huge effects on plant and animal life, from mass extinctions to migrations, to changes in agricultural practice that will affect global human food supply and production. However, most of us do not have a tangible understanding of the specific kinds of impacts possible. Let your course laboratory module be a place where students can explore and discover some of these effects while learning some key fundamental skills and techniques.

This module takes about 6 weeks to complete using a typical semester calendar. You could extend it to 7 or 8 weeks if you allow the groups to repeat/extend their experiments to more closely mimic how actual experimental science is done. Your overall course could comprise two different laboratory modules.

Weeks 1 and 2: Observing Plant Cells and Measuring Plant Primary Productivity – Two Laboratory Weeks

These procedures are very similar to what has been done for decades in your introductory biology course laboratory. The key difference is the conceptual organization. Rather than having the students simply learn the techniques because they are "what every biologist should know," present the skills as the toolkit they will need to conduct the experiments that address the conceptual questions posed by the module. If you think back to when you first began your career as a researcher, you likely entered a laboratory with a guiding question or area of exploration. You first learned the techniques you needed to address that research question and then designed and carried out specific experiments using those techniques and approaches. Most often, you needed a little bit of hands-on training, coupled with readings to figure out what experimental questions to ask. With the guidance of the head of that research laboratory, you learned how to do science by doing science using techniques and skills available in that laboratory. It's really not the particular techniques that matter, but rather it's the conceptual approach to using them that *is* the science.

Spend a little time at the beginning of the laboratory module introducing the guiding question or overall experimental goal. The title of the module says it all, but it's important to provide a little more background framing, the "hook" to get your students curious. What is known about this question? What work has been done before (preferably talk about some experimental studies that used some of the techniques your students will use themselves). What questions remain? If you have an experimental research paper for your students to read and then use that as a basis for discussion to help provide background and context, so much the better. Then, your students will have the internal curiosity and motivation to learn the techniques and at the same time appreciate how these approaches can be used for further exploration and discovery.

In the laboratory module I've developed here, I'm using *Elodea* as a model organism to explore the effects of climate change such as temperature or decreasing pH on primary productivity. You could also use Duckweed, which is another easy-to-work-with organism, or even fresh spinach leaves from the local supermarket. In any case, you want your class to understand what is meant by "model organism" and why they are useful.

Week 1: Examining Properties of *Elodea* that Make It a Good Model System for Evaluating Aspects of Climate Change on Primary Productivity

The following are the major goals and questions students will work on during this initial laboratory session.

(1) Observation of chloroplasts and plant cell structure of the model organism *Elodea*.
  (a) Why is *Elodea* a good model organism for studying plant function?
  (b) What characteristics does *Elodea* share with most plants? What are some aspects unique to *Elodea*?
  (c) What kinds of experimental questions can we address using *Elodea* as a model organism? What kinds of questions cannot be addressed using *Elodea*?

  For this part of the laboratory session, students will learn how to prepare living specimens for observation under a light microscope and will also learn how to use a microscope. Pretty standard stuff, but with an organizing theme and discovery-based goal.

(2) Understanding the molecules important for plant primary productivity: chlorophyll extraction.
  (a) How many pigment molecules participate in photosynthesis in *Elodea*?
  (b) Can you identify particular pigments using this extraction technique?
  (c) How might you use this technique to quantify levels of pigment molecules in different plants or in plants exposed to different environmental conditions?

  This simple extraction can be performed and used as a way to talk about the pigment molecules involved in photosynthesis. The crude extract can also be used if you measure primary productivity using a spectrophotometric assay the following week. This will give your students a second week practicing the techniques. There are various buffers, all kept chilled on ice, in which to grind up either fresh leaf discs or frozen leaf discs of known area. Two key components of the buffers are insoluble polyvinylpolypyrrolidone (PVP) and acid-washed sand. Then, the homogenized material is filtered through Miracloth and spun in a centrifuge. The resulting supernatant can be used in paper chromatography to visualize different photopigments and can also be used in spectrophotometric assays of enzymes such as Rubisco, PEP-CK, or NADP-ME. (See Further Reading at the end of the chapter for articles and references on these enzyme assays).

Week 2: Measuring Primary Productivity in *Elodea*

During this laboratory session, students learn the assay(s) that you want them to employ during their independent experimental investigations. They can address the following questions:

(a) What level of productivity do you measure in untreated *Elodea* extracts?
(b) How reliable are your measurements?
(c) What conditions might alter the apparent photosynthetic activity?
(d) What sorts of controls might you need to include when using this technique to measure primary productivity?

The students learn skills and techniques such as timed assay, close observation, and how to organize data for statistical analysis. They also practice important skills such as making serial liquid dilutions, measuring small accurate volumes of liquid, and the like. If you introduced the students to the crude extract preparation technique the previous week, student groups practice this technique further. There are a number of straightforward techniques for measuring primary productivity in leaf extracts or leaf segments. I have included the original sources in the "Further Reading" at the end of this chapter. The important aspect in this case is not the snazziness of the technique (in fact, simpler techniques are better so that the students focus on what photosynthesis is rather than on a complex set of instructions), but rather the application of the scientific process using this technique or skill.

The goal is for your students to learn how to quantify a process and how to analyze and present quantitative data. They will use this technique in their independent experiments. This laboratory period gives them practice using the technique and also introduces them to key quantitative approaches used in many different subfields of biology.

Simple Assays of Photosynthesis/Primary Productivity

(1) Biochemical assays of photosynthesis
    Students perform the crude leaf extract procedure that they used the previous week to measure Rubisco activity (Suzuki et al., 1987, see Further Reading) or other enzymes such as NADP-ME or PEP-CK (Ku 1991, see Further Reading). Product production can be measured

in a timed absorbance spectrophotometric assay at 340 nm and represented as absorbance/mg leaf sample/time. Students graph their results as a measure of photosynthetic enzyme activity. For independent experiments using this assay method, students could look at rates of reaction in extracts that are at various pHs or various temperatures.

For *Elodea* (or other simple aquatic plants such as *Cabomba*), it is also possible to measure the pH of the surrounding liquid containing a known quantity of *Elodea* leaves. Recording the pH reduction every 10 min for an hour will yield a quantitative measure of the rate of net photosynthesis (minus respiration) that correlates with primary productivity.

(2) Leaf Disc Oxygen Bubble Assay

Developed by Brad Williamson and available on the Internet (see Further Reading), this is a very simple assay that also really gets the point of photosynthesis across to your students. Students prepare flat discs of the same measured diameter and thickness using smooth, bright green leaves such as spinach or Pokeweed. You could also use individual duckweed leaves, but students would have to be sure to have the same overall area of material represented in their different treatment groups if they use this assay for independent experiments.

The principle is simple, but the biology behind it is a great example of energy transformation and metabolic pathways at the level of cells and whole leaves (tissue level). As oxygen is produced and accumulates in the extracellular spaces between plant cells in a leaf disc, the disc will float because of the lower density of oxygen than water. To set up the assay, students need to replace the oxygen with a water-based bicarbonate solution, the bicarbonate then serving as a consistent source of carbon dioxide for the photosynthesis reaction. The prepared leaf discs begin the assay at the bottom of a tube, and, as oxygen is produced within the leaf discs, they float to the surface. Students can measure the rate of photosynthesis (minus the rate of oxygen utilization due to respiration) by measuring the time for all the discs to float or by measuring the number of floating discs per unit of time. When all the leaf discs have floated to the surface, you can cover the set up and measure respiration rate as the rate at which the leaf discs sink. Students can quantify their findings, graph their results, and do statistics if different experimental conditions are tested.

Because this assay is so simple, some of your students might have performed it in AP Biology in high school. You can enhance the level of sophistication by having students design and carry out independent experiments using this assay as a measurement tool. You can remind your students that just because they might have learned a particular skill or technique in the past (such as using a micropipet or an electrobalance), it does not mean that the technique/skill is not useful at more advanced levels of biological experimentation. You can have your student groups choose which assay they prefer. Then in a class discussion or during student presentations, you can discuss the advantages and disadvantages of each assay technique.

Week 3: Designing Independent Experiments to Explore the Effects of Climate Change on Primary Productivity in Green Plants

This laboratory session focuses on designing an experiment to explore aspects of climate change or fertilizer excess on primary productivity using *Elodea* or Duckweed leaves or other plants as model systems. Students work in groups to design an experiment that includes appropriate controls and adequate sample size and that considers how they will analyze their data. Students then consult with you about their designs and consider revisions you suggest. They need to think about things such as the supplies they need, any solutions they need to make, whether they need instruction in the use of a pH meter, or balances. This way of learning-by-doing some basic laboratory techniques such as liquid measures and dilutions and operation of a pH meter and balance will be much more relevant to the students than a cook-book laboratory where they measure things such as pennies or erasers just to learn how to operate the instrument. They are putting the technique to use, rather than going through the motions to satisfy a laboratory assignment. They are also learning how to work together as a research team. Your role during this laboratory session is as a consultant. You can also use this time to train small groups of students on how to use the statistical tests that they plan to use or other general laboratory skills they may need help with.

Week 4 and 5: Student-designed Discovery-based Experiments and Data Analysis

In this laboratory session (and the next if you have time in your academic schedule to permit multiple experiments), student groups conduct their independent experiments, gather data, and plan out their analysis and presentation. During the laboratory period, you can circulate among the groups, helping them trouble-shoot problems that arise and talking with them about their results and how best to analyze them. If there is time in the laboratory period, you can talk with the groups as a whole about the kinds of statistics they would use to analyze their data. Have your groups complete their data analysis and plan for any repetition or revision to their experiment as a homework assignment. Just learning a technique and collecting data are only part of the process of doing a scientific experiment. Analyzing the data and critically evaluating the data and how the study might be improved or whether it needs repeating are also crucial aspects to doing/learning science. When students have their own data, they will learn the analysis techniques much better than simply having a simulated data set or a bunch of weights of coins or objects.

Week 6: Field Observations of Plant Communities in Areas Exposed to Fertilizer Run-off or Other Human Activity such as Road Salt Application in the Winter

This laboratory exercise could be a field trip experience if your area allows. Or, alternatively, students could view a film or series of online videos. An essential aspect of this experience would be to link the material to both this independent experiment experience and the course material. Students can discuss about how to measure the health of plants growing under different conditions. This exercise can greatly enhance the relevance of the laboratory work and the course material to current issues confronting our society or our local area. If you are viewing videos, you could couple this with student group presentations of their data and hold a class discussion about the results and how they relate to findings in the field.

Assessments

Skills Mastery

Student progress in quantitative reasoning can be evaluated by having your students prepare figures and graphs, and by having them conduct a statistical analysis of their data. Evidence such as the sample size, the choice of statistical tests, comparisons with appropriate control groups, and the quality of the figures can all be inferred from a good figure and figure legend. Assigning problem sets or requiring that students write a self-reflection statement about their quantitative reasoning process provide additional practice and opportunities for assessment.

Concept Mastery and Links to the Classroom

Often, introductory biology students fail to appreciate the relationship between the laboratory and the classroom. Perhaps this is because the timing of the two gets rapidly out of sync. A 6-week long laboratory experience takes half a semester, which could correspond to as many as 36 classroom sessions. One way to emphasize the link between them is to refer to the laboratory projects during class and to refer to the class during laboratory. Another way is to include the laboratory project material on classroom assessments, such as quizzes or examinations. You could also assign short essays in your laboratory to give your students practice writing and drawing connections between the laboratory projects and the fundamental biological concepts.

To assess the overall mastery of the concepts of the laboratory module and to give students practice communicating science, create a blog or website that is your course "online journal." Students can submit their projects to this site for other students in the class to read. The submission could be as simple as a figure and descriptive figure legend or could be a laboratory paper written in the form of a scientific paper. The submission could include explicit connections to class material. You could ask the other students in the course to read one or more and provide peer review or comments online.

Taking Stock

This module allows both the development of many key skills such as microscopy, observation, spectrophotometry or assay measurements, macromolecule separation/extraction, experimental design, quantitative analysis, graphing, statistics, and field measurement techniques and at the same time the development of abilities to convert a question of interest into a testable experiment and to analyze data and present them. Finally, an organizational scheme such as this allows students a chance for both self-discovery (learning new things that are already known by others) and real discovery (exploring impacts of human activity on plant function).

A number of the core concepts and levels of biological organization are represented in this 6-week laboratory module: structure/function (leaf anatomy and microscopy); energy transformation and pathways (leaf disc assay or enzyme assay); and systems.

Module 2: How Does Antibiotic Resistance Develop?

This organizing idea, first mentioned in Chapter 2, focuses on processes of evolution, particularly selection mechanisms such as natural selection and artificial selection. It also emphasizes information storage and transfer, as well as structure–function relationships. Many students taking introductory level biology courses are interested in human health. In addition, a key area identified by scientists and educators for college-level students is the issue of human health and disease. This module addresses those areas and is also quite relevant to students' everyday experiences.

This laboratory module teaches important laboratory skills such as sterile technique, bacterial plate assays, growth assays using spectrophotometers, PCR, DNA or protein gel electrophoresis, DNA restriction digests, and mutation analysis. This module also reinforces concepts such as selection pressures and micro-evolutionary change across generations. Students discover and explore how strong versus weak selection affects population growth, dynamics, and evolutionary change. Student groups examine how the evolution of antibiotic resistance is influenced by multiple selection pressures operating at once, modeling real-world scenarios.

Week 1: Observing cell division; Measuring bacterial Growth and Introduction to Sterile Techniques

Goals and questions:

(1) How are prokaryotic cells similar and different from eukaryotic cells?
(2) How does eukaryotic cell reproduction (i.e., mitosis and meiosis) differ from prokaryotic cell division? How are they similar?
(3) How can a bacterial growth assay be used to examine antibiotic resistance?

Students need first-hand experience comparing prokaryotic and eukaryotic mechanisms of cell division and reproduction. This entrée to the module gives students additional practice with close observation and further develops their microscope skills. Students also learn sterile technique and, when coupled with Module 1, additional macromolecule extraction/analysis techniques.

(a) Observation of mitosis and other forms of cell division and growth/reproduction – comparison between prokaryotes and eukaryotes

Before the laboratory session, students view a number of videos and images available online that compare prokaryotic and eukaryotic cells and cell division (see Further Reading).

Students observe prokaryotic cells by looking at either cyanobacteria or a small sample of yogurt with live bacterial cultures, diluted in water. To compare with eukaryotic cells and to observe mitosis, students prepare onion root tip slides in a simple procedure that takes about 20 min.

Have on hand for this laboratory: yogurt or cyanobacteria and some onions with actively growing root tips. There are a number of different published materials describing how to prepare microscope slides for viewing prokaryotic and eukaryotic cells and for visualizing mitotic stages using onion root tips. For the root tip preparation, you need to have actively growing roots from placing an onion bulb in water to allow root growth. The root tips are cut, fixed in acetic alcohol or other solution, mechanically disrupted, stained with aceto-orecein stain, and then coverslipped and viewed (see Further Reading). If students prepare these slides themselves, they will get more practice with microscopy and specimen preparation; however, if you have time and resource constraints, commercially prepared slides are readily available.

(b) Two different assays of bacterial growth:

A crucial companion to the microscope observations is the quantitative measurement of bacterial cell growth. A growth assay can be used by students in subsequent weeks for their antibiotic resistance projects.

Bacteria growing in liquid culture cause the liquid to become turbid. The turbidity is measured with a spectrophotometer (at 600 nm), and the rate of increase of turbidity, measured as the absorbance or optical density (OD), is an accurate measure of culture growth rate.

Students learn about how to inoculate a flask of bacterial growth medium, such as LB broth, with a standard laboratory strain of *Escherichia coli* and incubate the flask in a shaker incubator or water bath. Students remove 1 ml of the solution every 20 min (the generation time for many common laboratory bacteria is about 20 min) and place it in a cuvette to measure OD at a wavelength of 600 nm (this measurement corresponds to light scattered due to increased cell concentration). From the resulting graph of OD versus time (minutes), students determine the growth rate and the generation time. Later, in student-designed experiments, students can compare growth rates of cultures exposed to different concentrations of antibiotics or different durations of exposure to antibiotics (development of resistance).

Week 2: Plate Assay or Turbidity Measurements to Examine Antibiotic Resistance, Design of Selection Experiments

Goals and objectives:

- How do antibiotics affect bacterial growth?

- Design an experiment to examine the evolution of antibiotic resistance in bacteria.

When selecting a measurement assay, in this case either the liquid culture growth assay or the plate assay for zone of antibiotic inhibition, what strengths and weaknesses do you need to keep in mind for each? How does the choice of assay influence your experimental design?

(a) Pilot Experiment for Planning Independent Project

Have your students perform a pilot experiment using each assay technique. You can either provide students with bacterial cultures already grown in broth or have students grow their own cultures. If the latter, set up the initial cultures the previous week so that they are ready for this laboratory session. For example, students can swab their mouths or skin surface and can streak the samples on agar plates to obtain their own cultures. Single colony isolates from their plates can be grown in liquid culture for use in the selection experiment.

(b) Effects of antibiotic on bacterial growth using the liquid culture assay

Students set up two sets of six to ten tubes (9 ml volume) of LB Broth. To one set add a predetermined concentration of antibiotic of their choice from a selection. Students take 1 ml of their first antibiotic solution and add it to each of two tubes, labeled 1A (no antibiotic) or 1B (antibiotic). Take 1 ml from each of these and add to a second set of tubes 0.1 A or 0.1B. Continue these serial dilutions four or five times, add the same amount of bacterial culture to each, and then incubate all of these cultures at 37 °C for different lengths of time, removing 1 ml at 20′ intervals to measure turbidity at $OD_{600\ nm}$. The slope of the best fit line from a plot of OD_{600} versus time for each tube represents the growth rate.

(c) Measuring the effects of antibiotic on bacterial growth using a plate assay.

For this assay, students spread 100 µl of their liquid culture on each of two petri plates that are filled with a growth medium (LB or TSB are commonly used). They let the cultures soak into the plates and then apply small antibiotic disks to equidistant quadrants of each plate (including some antibiotic-free disks for control) and incubate at 37 °C overnight. The next day, students measure the zone of inhibition (Figure 3.2), or they can move the plates to the refrigerator until the next laboratory period.

(d) Experimental Design Consultations

During this laboratory period, visit the students groups to discuss their experimental designs. Alternatively, you could have your groups

Figure 3.2   Zones of growth inhibition surround different antibiotic disks. (CDC Public Health Image Library. Image credit: CDC/Dr. JJ Farmer (PHIL#3031) 1978. Upload date: 8 March 2006 by Marco Tolo. Permission PD-USGOV-HHS-CDC.)

sign up for consultation sessions either later during the laboratory period or at some point during the week. During the consultations, discuss things such as the procedure they intend to follow, controls, sample size, statistics they plan to use, graphical or other data analysis, timing of their experiment, and so forth. If you have constraints on supplies, you can provide students with a selection of antibiotics on disks for a simple plate assay technique. Students can also be given a stock solution of a number of common antibiotics, such as ampicillin or streptomycin or tetracycline. Students can prepare a dilution series with one or more antibiotic and spread them on agar plates. A handy way to measure lots of different concentrations would be to use micro-titer plates or multiwell culture plates with solid agar. Students can then work with multiple groups or replicates in a single plate. A third possible set up would be to have students take individual colonies from their bacterial plates and inoculate small (5 ml) liquid cultures containing different concentrations of antibiotic. After incubation at 37 °C overnight, cultures can be kept in the refrigerator until measured during the next laboratory period. Make sure your students write up their experimental plan and provide you with a copy. This will give them a chance to review their plans and will also give them practice writing and articulating an experimental design. It also gives you a copy of the plans to help you provide any materials they might need.

Weeks 3–5: Independent Experiments Examining Antibiotic Resistance

This laboratory module gives students experience with experimental design and data analysis, as well as sterile technique and microbiological techniques. They also get a chance to repeat and practice what they are learning by being able to design and conduct an experiment more than once. For the next 3 weeks, your student groups come into the laboratory and set up their experiments using their experimental designs as a guide. They count colonies or zones of inhibition or OD_{600nm}, gathering data. If your students conducted their resistance experiments in liquid culture, they can measure growth rates by taking an aliquot from each culture and growing it in liquid culture, measuring turbidity over time, to generate growth curves for each condition they examined. If your students conducted their experiments on solid agar plates, they can count the number of colonies on each plate (they should have been careful to add the same amount of bacteria to each plate the previous week) or the size of zones of inhibition. Of course, the first week the groups set up their experiments for the first time. Most likely they will have lots of questions and will need guidance with equipment, with serial dilution techniques, and with sterile technique. Some may realize that they need to make adjustments to their experimental designs because of time or equipment constraints. Your laboratory room will be a chaotic, action-packed place humming with student conversation.

To have your students continue to exert "selection pressure," have them take colonies that appear on the lowest concentration antibiotic plates and grow them in liquid culture with the same concentration. After about an hour of growth at 37 °C, they can streak plates with that concentration and slightly higher concentrations. You can have students repeat these steps a number of times to expose several generations of bacteria to the same conditions. Each week, students can "select" surviving colonies in the antibiotic zones of the plates, exerting an artificial selection pressure. Students can manipulate the number of generations they exert selection, the concentration of the antibiotic that they use in the selection paradigm, and other ideas that they may have come up with during their experimental design session.

An amazing transformation takes place with the repetition of an ongoing independent project. There are fewer questions about technique or procedure and more about the implications of the results or whether the sample size is sufficient, or how the results fit with other work in this area. A smooth order of actions develops. The groups begin to splinter into efficient subdivisions. One might be an expert plate-spreader. One might be gathering data, while another is setting up the next experiment. They work as a team, conferring with each other, rather than you, as they begin to see results emerge. At this point, they can begin to analyze their data and figure out how best to present their findings.

Week 6–7: Continued Experiments if Time Permits

There are a number of genes that are commonly associated with antibiotic resistance that can be assayed using PCR of extracted DNA. After students exert selection pressure on their bacterial colonies multiple times (say, 3 or 4), students can select resistant colonies, grow up the bacteria in liquid culture in the presence of the antibiotic, and then extract DNA for PCR analysis. If you choose to add in a couple of weeks of macromolecule-level analysis to their projects, be sure to have your student groups select resistant colonies and control colonies the week before this, so that liquid cultures are available for DNA extraction and PCR analysis. There are a number of cost-efficient kits available, as well as published PCR primer sequences (see Further Reading) for rapid extraction and analysis. The first week is spent extracting and setting up the PCR; the second week students use gel electrophoresis to separate and analyze their PCR fragments. This last week can also be used for finalization of figures and data analysis for the assignment associated with this module.

Assessments

Culminating assignments for this laboratory module could include a laboratory report written in the style of a scientific manuscript, or an oral or poster presentation in the style of a research conference, professional quality figures, and class discussion about natural/artificial selection, evolution, antibiotic resistance, and public health.

Taking Stock

This laboratory module builds on microscopy and other fundamental laboratory skills. It also supports core concepts of structure–function, information transfer, and evolution. It can support classroom topics such as mitosis/meiosis, mutation, growth and metabolism, prokaryotic versus eukaryotic cell structure and function, evolution by natural or artificial selection, and genetic variation.

Module 3: Self-Discovery Explorations of Human Diseases Caused by Single Nucleotide Polymorphisms

Goals:

(1) Learn how to use bioinformatics and web-based tools to conduct research into human disease.

(2) Integrate microscopic observations of normal tissues with literature-based information about disease effects on tissue morphology and function.

(3) Integrate microscopic and bioinformatics investigation with a PCR-based single nucleotide polymorphism (SNP) study of a human gene and extrapolation of this approach to a genomic study of human disease genes.

This module focuses on literature-based, large database-focused self-discovery of a student-chosen human disease that has a genetic component. This important exploratory work is complemented by learning how to extract DNA (or to continue to practice this skill if you included this technique in a previous laboratory module), how to perform a restriction enzyme digest to generate fragments of DNA, and how to do polymerase chain reaction to identify SNPs. Students explore a disease on multiple levels of biological organization using the published scientific literature. In the laboratory, students examine key tissues and organ function using a microscope, in a format similar to that in many published introductory biology laboratory manuals. The emphasis is on major concepts such as structure/function, cell and tissue organization, and comparative and evolutionary processes.

Cancer at a cellular level is unregulated growth and inability of the cells to respond to growth inhibition signals provided by tissues. Cancer results from a finite accumulation of mutations in genes that are involved in the regulation of cell growth. Many of these mutations are single nucleotide substitutions that affect how the genes work. In addition, many growth regulatory genes in the human population vary by single nucleotides, so-called SNPs, which underlie variations in how cancer develops, in the susceptibility to developing cancer, and in how patients with cancer respond to treatments. Students can discover and learn about SNPs, genetic variation, evolution, information flow, cell/tissue communication pathways, and much more. In the laboratory, students can observe different tissues and examine the ways that cell growth is restrained. If you did not use Module 2 mentioned previously, you could include observations of mitosis, meiosis, and cell growth in this module. Students compare this growth with growth of immortalized tissue culture cells. They can compare microscope slides of tissues in different animal and plant taxa to evaluate similarities and differences in tissue organization. They could also explore tissues and organs at the organismal level via dissection.

There are a number of other diseases caused by SNPs, including cystic fibrosis, sickle-cell anemia, Huntington's disease, myotonic dystrophy type 1, Neurofibromatosis Type 1, Rett's syndrome, Crohn's disease, and Hemophilia A. Students learn how to do a SNP analysis, using a non-disease-causing SNP in the human population, such as for TASR 38, a gene that codes for phenylthiocarbamide (PTC) tasting ability, or for cdk3, a gene for which a student laboratory module has already been developed (see Further Reading).

The goal is not to survey animal/plant taxa, but rather to understand cell/tissue/organ structural and functional organization with the aim of understanding how cancer interferes with that structure and function. Students could choose to focus their investigations on particular subsets of tissues or organs, those that are affected by the type of cancer they are investigating.

Week 1: Student Investigation Specific Aims and Goals – Use of Bioinformatics to Explore Genetic Diseases Associated with SNPs

Goals and activities:

(1) Have students fill out a guided inquiry worksheet. Ask them to provide some information about their gene of interest: what chromosome it is on, tissue expression, gene product function, role in the disease (if they can find this), what the actual SNP is, and whether there is a diagnostic test for the SNP.
(2) Have students develop a bibliography of several research articles that address the SNP involved in their chosen disease.
(3) Have student propose a series of experiments or tests to explore the SNP-based disease further.

Students use the experimental literature to select a SNP-involved disease that they wish to explore. They outline their objectives, taking care to include investigation at multiple levels of biological organization. Students will examine healthy cells/tissues that are affected by the disease of interest. The students select a gene from a list you provide and learn about how that gene is involved in the disease progression, symptomatology, or susceptibility.

In this laboratory session (adapted from Banta et al; see Further Reading), students work with computers to investigate their gene using the NCBI Bioinformatics site. Students learn how to access the vast information warehouses at the NCBI, including how to find genes associated with human disease, how to search for genetic variation in the human

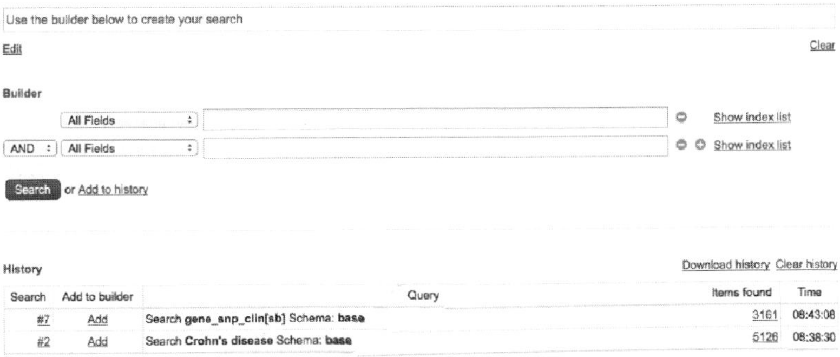

Figure 3.3   Excerpt from a bioinformatic search of genes involved in Crohn's disease.

population for those genes, and how to link published experimental literature with the genomic information in the databases.

As an example, I've chosen Crohn's disease, a recent example of a SNP-based disorder. Student begin their search for SNPs in genes associated with Crohn's disease using the "Gene" database with advanced search filters; see Figure 3.3.

By adding the two search terms, students get a list of SNPs associated with their disease. They then choose one or a few of these genes for further exploration about their functions in particular tissues and about the genetic mutations associated with Crohn's disease.

These databases are huge, and the interfaces change regularly as the databases are expanded and revised to accommodate new data-mining strategies. It is really important to provide some guiding instructions that are clear and that work for the diseases that your students select. There is so much information that students can quickly become overwhelmed and lost in the dense thicket of this massive information warehouse. Check out the Further Reading section of this chapter for links to guidance for students.

What students will see no matter what disease-associated gene they explore is that often the gene product has more than one function and is expressed in multiple tissues. This is a key fundamental concept in biology, that many genes are pleiotropic.

Through their bioinformatics investigation, students identify a SNP gene sequence that could be analyzed using PCR. Students develop a plan for using PCR to determine the disease genotype of individuals. Then, in the following week's laboratory, students will use that strategy to examine their own DNA genotype for a non-disease-causing SNP.

Weeks 2 and 3: SNP Analysis for TASR 38 or cdk3 Using Polymerase Chain Reaction

Goals and activities:

Students extract DNA by swabbing their own cheek cells. This technique is used both for medical diagnostics and for forensic analysis, so it's a very useful skill to acquire. DNA can either be extracted using a number of cost-effective and convenient DNA testing kits or be extracted using solutions you make yourself.

After DNA extraction, the pairs or small groups of students set up a PCR reaction to amplify a small DNA fragment that contains one SNP. The PCR reaction can run overnight and the tubes moved to a freezer for storage until the next laboratory period. The data set for this analysis is the entire class combined together. You can compute the frequency of the TASR38 gene or cdk3 SNP in your class and compare this frequency with that of the broader population. Comparing these gene frequencies enables an important discussion of sampling issues. If you choose TASR 38 as the gene to examine, you can have your students perform a simple PTC taste test. Many students might have done this test in earlier grades in school. PTC is a bitter compound that about 75% of the human population can taste. Twenty-five percent contains a SNP in the TAS2R38 gene that renders the taste buds unreactive to PTC.

The next laboratory period the students perform a restriction enzyme digest that will yield different sized fragments on the basis of the presence or absence of the SNP. The resulting DNA fragments are run out on an agarose gel and students genotype themselves. The restriction fragment pattern can reveal whether a student is heterozygous or homozygous for the SNP. The intensity of the bitter perception is correlated with the number of copies of functional TAS2R38 gene.

This module links what students are learning about information storage and transfer in the form of DNA and genes and mutations with structure and function relationships. The module also integrates across multiple levels of biological organization, from molecule to tissue to organism to population. The use of students' own DNA, coupled with the taste perception test and the use of large online databases with genetic and genomic information, makes the learning relevant and personal.

Assessment Ideas

Because the students' work is self-determined in terms of their literature-based investigation of SNP-caused diseases, you will want your assessments not only to probe mastery of core concepts, but also to give students practice disseminating scientific information and content.

Table 3.2   Discovery-Based Laboratory Modules

Lab Topic	Techniques and Skills	Concepts Addressed
Module 1: Human Impacts on primary productivity		
Week 1: *Elodea* Structure and Function	1. Microscopy 2. Molecular Extraction (chlorophyll)	1. Cell structure/function 2. Macromolecule structure/function 3. Plant structure/function
Week 2: Measuring primary productivity	1. Macromolecule extraction 2. Enzyme assay or energy production assay (leaf disc) 3. Experimental design	1. Energy transformation 2. Enzyme function 3. Photosynthesis and respiration 4. Green plants
Week 3: Student-designed independent experiments	1. Experimental design 2. Assay practice 3. Quantitative data analysis	1. Scientific process 2. Working with data sets 3. Quantitative Reasoning 4. Statistics
Week 4: Student-designed experiments II and data analysis	1. Quantitative data analysis 2. Interpretation and presentation of data	1. Scientific process 2. Working with data sets 3. Writing and figure preparation
Week 5: Field trip to observe human effects on green plants	1. Field-based measurements 2. Categorizing biodiversity	1. Plant diversity 2. Ecosystem structure and function
Week 6: Summative Event	1. Poster or oral presentation	1. Writing, speaking Science
Module 2: Antibiotic Resistance as Selection for Evolutionary Change		
Week 1: Comparison of prokaryotic and eukaryotic cell structure and reproduction	1. Microscopy 2. Onion root tip mitosis 3. Turbidity assay	1. Biodiversity: Prokaryotes versus Eukaryotes 2. Cellular level structure/function 3. Mitosis 4. Cell division 5. Information transfer
Week 2: Experimental design and Plate assay techniques- Experiment I	1. Sterile technique 2. Serial dilution technique 3. Experimental design 4. Scientific process	1. Scientific process 2. Micro-evolutionary change 3. Genotype to phenotype 4. Artificial Selection

(Continued)

Table 3.2   *(Continued)*

Lab Topic	Techniques and Skills	Concepts Addressed
Week 3: Experiments II	Same	Same
Week 4: Experiments III	Same	Same
Week 5: Summative Event	1. Poster or oral presentation	1. Writing and Speaking Science 2. Integration and synthesis
Module 3: Human Disease and SNPs		
Week 1 : Bioinformatic analysis	1. Use of computers to mine large public data sets and databases	1. Genetic variation 2. Mutations, DNA, replication 3. Structure/function 4. Evolution
Week 2: SNPs from cheek cells and PCR	1. Cell extraction techniques 2. PCR 3. Micropipetting and liquid measures	1. DNA variation/mutations 2. DNA replication and PCR
Week 3: DNA gel electrophoresis	1. DNA as genetic material	1. Information transfer 2. Mutations 3. Properties of macromolecules
Week 4: Summative Event	1. Poster or Oral Presentation 2. Experimental Design Assignment	1. Science communication 2. Integration and Synthesis

Having your student groups give short oral presentations is one way to accomplish these assessment goals, but this format will work only for small classes of perhaps no more than 25–30 students. You could also hold a student research poster symposium for middle-sized courses of probably no more than 50 students. What about very large classes? You might consider a poster symposium for selections among applicants. The rest might present their work in the form of a blog post. You could have students be required to view and comment on a subset of the blog posts.

Let's put these modules together and see what we have come up with. Depending on how many weeks you have available and how many weeks you make each module sketched out previously, you might have time for two of the aforementioned modules as written. Table 3.2 summarizes the three example modules presented in this chapter.

Summary

In this chapter, I have shown you how to take a standard introductory biology laboratory series and rearrange and combine the various techniques into modules that are connected by a theme. Although not every aspect of the "old" laboratory sequence was addressed, in particular the surveys of different organism groupings, many of the fundamental techniques and approaches are still there. Sounds good on paper, but will it really work?

Further Reading

1. Rissing, S.W. and Cogan, J.G. 2009. Can an inquiry approach improve college student learning in a teaching laboratory? CBE-LSE 8: 55–61.
2. DebBurman, S.K. 2002. Learning how scientists work: experiential research projects to promote cell biology learning and scientific process skills. CBE 1: 154–172.
3. Brewer, C. and Smith, D. (eds.). 2011. Vision and Change in Undergraduate Biology Education: A Call to Action. American Association for the Advancement of Science. ISBN#: 978-0-87168-741-8. http://www.visionandchange.org/
4. Vodopich, D.S. and Moore, R. (2014) Biology Laboratory Manual, 10th Edition; McGraw-Hill Companies, Inc. New York, NY.
5. Exploring Photosynthesis with Fast Plants, Department of Plant Pathology, University of Wisconsin-Madison, 1991. wfp@fastplants.cells.wisc.edu
6. Fox, M., Gaynor, J.J. and Shillcock, J. 1998. Floating spinach disks- an uplifting demonstration of photosynthesis. J. Coll. Sci. Teach. 28(3): 210–212, http://ogobio.weebly.com/uploads/3/2/3/9/3239894/ap-_lab-photosynthesis_inquiry.pdf]
7. Suzuki, S., Nakamoto, H., Ku, M.S.B. and Edwards, G.E. 1987. Influence of leaf age on photosynthesis, enzyme activity and metabolite levels in wheat. Plant Physiol. 84: 1244–1248.
8. Ku, MSB, Wu, JR, Dai, ZY, Scott, RA, Chu, C, and Edwards, GE. 1991. Photosynthetic and photorespiratory characteristics of *Flaveria* species. Plant Physiol. 96: 518–528.
9. Student Learning Assessment: Options and Resources, 2nd Edition. Middle States Commission on Higher Education. 2008.
10. Haddix, P.L., Paulsen, E.T. and Werner, T.F. 2000. Measurement of mutation to antibiotic resistance: ampicillin resistance in *Serratia marcesens*. Bioscene, v26, 1-21.
11. Video of prokaryotic and eukaryotic cells and division: http://www.youtube.com/watch?v=uU00tg98Jjw
12. Pishva, E., Havaei, S.A., Arsalani, F., Narimani, T., Azimian, A., Akbari, M. 2013. Detection of methicillin-resistance gene in *Staphylococcus epidermidis* strains isolated from patients in Al-Zahra Hospital using polymerase chain reaction and minimum inhibitory concentration methods. Adv. Biomedical Res. 2(2): 1–5.

13. TAS2R38 gene in OMIM http://www.ncbi.nlm.nih.gov/entrez/dispomim.cgi ?id=607751

14. Wooding, S. 2006. Phenylthiocarbamide: a 75-year adventure in genetics and natural selection. Genetics 172: 2015–2023: http://www.pubmedcentral. nih.gov/articlerender.fcgi?artid=1456409

15. Banta, L.M., Crespi, E.J., Nehm, R.H., Schwarz, J.A., Singer, S., Manduca, C.A., Bush, E.C., Collins, E., Constance, C.M., Dean, D., Esteban, D., Fox, S., McDaris, J., Paul, C.A., Quinan, G., Raley-Susman, K.M., Smith, M.L., Wallace, C.S., Withers, G.S., Caporale, L., (2012). Integrating Genomics Research throughout the Undergraduate Curriculum: A collection of inquiry-based genomics lab modules. CBE: Life Sci. Educ., 11, 1–5, Fall 2012. (modules available at: http://serc.carleton.edu/genomics/units/snp.html)

4 The Constraints and Realities of Discovery-Based Laboratories

The ideas in the previous chapter sound so good on paper. Students learning relevant techniques, getting multiple opportunities to practice those techniques and to apply those techniques and skills to relevant experimental ideas that they generate themselves, and exploring current topics of interest such as climate change effects on organismal activity or the evolution of antibiotic resistance. If these laboratories are so easy and fun, why haven't introductory biology or life sciences courses been structured that way all along? Time, money, and resources are all limited. Also, face it, we faculty don't know everything about all life sciences subfields because we are all mostly specialists by training. And, to top it off, students know nothing about taking a question and turning it into a feasible experiment that can be done in 3 or 4 hours on a weekday afternoon. Do all these limitations mean that it's not possible? No, they just mean you have to plan things carefully. Let's take these constraints one at a time.

Instructor Expertise

When I read through laboratory experiences that have been developed by colleagues across the country, my initial reaction is excitement. What a great idea! I'd love to incorporate it into my own course. Then, the excitement quickly turns to anxiety. Wait a minute! I know nothing about: *Elodea*, Duckweed, *Arabidopsis*, *Drosophila*, field biology, molecular biology … …. This list goes on and on, and pretty soon I've decided it's easier to just go with what's already been tried and done successfully in my own department or in my own experience. Innovation falls victim to fear once again.

Discovery-Based Learning in the Life Sciences, First Edition. Kathleen M. Susman.
© 2015 John Wiley & Sons, Inc. Published 2015 by John Wiley & Sons, Inc.

It turns out that it's not as hard as it appears. You can take an organism or system that you have used in your introductory courses before, learn a little bit more about it, and modify your existing laboratories to the structures described in Chapter 3. Let's say you previously had your students look at chloroplasts of *Elodea* wet mounts during the "learn to use a microscope and make wet mounts" laboratory. *Elodea* is actually a native freshwater plant that's found all over North America. There are a lot of experiments that students could design and carry out to explore the effects of temperature, water pH, nutrient excess, or herbicides on plant productivity. With a little bit of effort, you can convert your technique-focused introductory laboratories into discovery-based experiences.

You can even use your own learning experience to organize your students' discovery. For example, let's say I want to use the *Elodea* laboratory module in my introductory biology laboratory the way that I outlined in Chapter 3. However, I'm a neurobiologist with not much training in plant biology or ecology. Turn that ignorance into a plan for your students' discovery. Make explicit the process that I use to start a new project. What do I need to know?

(1) What is *Elodea*? Where is it found? How is it similar to or different from other plants? How does *Elodea* contribute to its native ecosystem?
(2) As an aquatic plant, what environmental factors might influence its growth and productivity? How might one measure growth? Productivity?
(3) Might the factors that affect *Elodea* also affect other plants?
(4) What other plants grow in the local area near my college? What plants grow in the greenhouse at my college?

The Internet and your colleagues in plant sciences and ecology can help you quickly get up to speed. Your own experiences learning about *Elodea* and answering the aforementioned questions can serve as the scaffold for your laboratory sequence and even your classroom content. Use the questions you yourself investigated as the basis for self-discovery for your students. In introductory biology courses, much of what we want our students to learn has already been discovered. It is through self-discovery that we learn, master, and remember. By linking the self-discovery with a scientific process in the laboratory, students can extend and deepen their knowledge through a series of curiosity-driven experiences. Asking questions such as "What if ... ," "What happens if ...." and then designing experimental approaches toward examining those questions will give the students a tangible and deeper understanding of the scientific process.

As you learn about your new subject, explicitly fold that self-discovery process into your class curriculum. Let's take a concrete example. To learn about a new subject, you would probably start by consulting summary sources such as textbooks, encyclopedias, and reputable Internet reference sources. Then, you might find a review article about primary productivity of aquatic green plants by conducting a literature search using your institutional library. For example, in browsing my library's collection of articles, I found this: "The impact of insecticides and herbicides on the biodiversity and productivity of aquatic communities," by Relyea, RA, 2005 (see Further Reading). Another article jumped out: "Environmental influences on aquatic plants in freshwater ecosystems," by Freedman and Lacoul, 2006 (see Further Reading). These articles provide you with some ideas of the kinds of experimental approaches that might work with your class. The articles also reveal the current state of the field, the unanswered questions that might give you and your students some ideas to explore. Why not replicate this self-educational journey with your students by assigning a related experimental paper for discussion or a methodological paper to help guide their thinking for discovery-based experiments in the laboratory? You could assign segments of the textbook or online readings to provide background and context leading up to the experimental paper or review article.

Perhaps the most important task you will face in developing or implementing the discovery-based laboratory modules is to test the ideas out in advance. By trying a few experimental ideas yourself, you will get a sense of how much time is needed to do each part of the module. You will also get an idea as to how feasible the independent experiment ideas are to carry out by students. Finding the time to try out your new laboratory modules is very difficult. Perhaps you might have an independent research student who might test your ideas as a research project. If you have an undergraduate student work on idea testing, though, you need more lead time before you can implement your ideas in your course. Sometimes, I haven't had the time to trouble-shoot ideas before the course begins. In those cases, my student groups do the trouble-shooting.

Time

Possibly the scarcest resource of all is time. This is a huge constraint. Both the time it takes you as an instructor to develop feasible laboratories for introductory biology and the time the students have available in the laboratory. By grouping the laboratory modules around the kinds of exercises you already have in place in your introductory biology curriculum, you can be more efficient in your laboratory development time investment. If you are new to developing laboratories for courses, reorganizing an existing set of exercises and grouping them under one or two different research question areas will get you started. This is what we did in Chapters 2 and 3. As you gain confidence in the process of developing new laboratories, you will find you can branch out with less of an outlay of time and effort. Nonetheless, it does take substantial time and effort to develop laboratory modules that work, that are exciting, and that actually help you achieve your curricular goals.

Preparation Time

If your department has a technical preparation staff for the introductory laboratories, then you likely have lots of help getting the laboratory ready for the students. Consider asking them or your student interns or graduate student TA's to do a practice run of a new laboratory module a few weeks before the module begins. This will help you to identify aspects of the written instructions that are not clear, as well as to finalize all the different materials that students might need to complete their experiments. When I develop new laboratory modules, I usually try them out in the early summer months or during the break between semesters. Although it's true I don't get paid for that time and effort, I am able to use my department course budget to do the trial run. I'd much rather know that my laboratory module is likely to be successful than not invest the time and energy in the trial run.

Student Time In and Out of the Laboratory

The issue of student time in laboratory is a critical one. If your course is organized into three times a week "lecture" and once a week laboratory for 2–4 hours, which is the standard across U.S. colleges and universities, you need to plan carefully what a student group can accomplish in a single laboratory period each week. The actual procedures should take no longer than 1 or 2 hours to perform, in order to allow ample time for student groups to talk together, to organize themselves, to carry out the experiments, and to clean up at the end. When students are conducting experiments, you do not want them to feel pressured for time. You do not want them looking at the clock and getting discouraged. There is no benefit to having laboratories that run past the time allotted. Students often have other activities, other courses, and lots of class work to get done.

Having students prepare in advance of laboratory will greatly improve the use of time during the laboratory period. Consider assigning "pre-laboratory" tasks that require the class to read and reflect in advance. There are a growing number of good instructional videos and simulations available on the Internet that you can assign your students view before coming to the laboratory (see Further Reading). You might consider making short demonstration videos yourself using your own teaching materials and environment. Once you are in the laboratory, keep your introductory remarks brief and focused on the overall goal of the laboratory session and any safety issues. Then, circulate among the student groups for more specific instructions and to answer the inevitable questions. I have found that students really can't pay attention to a long set of verbal instructions, the so-called "pre-laboratory lecture." Your long speech about the steps of the procedures or about the use of the equipment will go in one ear and out the other. It's not relevant information until they get started. Of course, it is very difficult to get to all your groups at once. If you have a student intern or TA, you can split the room into zones to circulate and answer questions and trouble-shoot problems that arise (much like the waitstaff at restaurants).

Student teams can increase the efficiency of a laboratory session. Sometimes, students need a little help in how to effectively divide up the work. I usually have students work in groups of four. Then, I suggest they might pair up within that group after the full group has brainstormed their experimental idea to divide up the tasks and procedures.

Another way to help increase the efficiency of your students' time in laboratory is to build in group consultations with you in advance of the experiment sessions, particularly before the independent projects begin. If your groups provide you with a written experimental plan and bring that with them to a consultation session with you, you can help your students revise their plans into something feasible for the two or three experimental sessions devoted to their independent projects. You can also help your students by providing pre-made stock solutions they plan to use or organisms ready to go. For example, if a student group wants to explore ways in which water pH affects productivity of *Elodea*, you can provide them with solutions at the pHs that they wish to examine. Or, if a student group wants to evaluate how *Elodea* productivity is affected by a longer term exposure to water pH, you can discuss with them how they might do this and have them meet either you or a student intern/TA to place *Elodea* in water of different pHs for the time they wish to examine. Have them do this outside of laboratory time so that their samples are ready for analysis during the laboratory session.

You can also put restrictions on the length of planned experiments. For example, in my intermediate-level Neuroscience course, a student group wanted to explore the effects of a particular chemical on the lifespan of the nematode *Caenorhabditis elegans*. A worthy and interesting pursuit, but one that would take at least a month to complete just a single experiment. Instead, I suggested they examine the shorter term effects of that chemical in animals exposed at different stages of life and provided them with nematode cultures of different ages. Teaching your students how to ask questions and develop feasible scientific approaches within the constraints of a real-world situation is a very important skill for practicing scientists.

Time for Class and Laboratory – the Schedule of Classes

Most introductory life sciences courses are organized into a shorter "lecture" period that might meet two or three times a week and a longer "laboratory" period that meets once a week for several hours. Some institutions offer separate lecture and laboratory sections (i.e., the students register for them as separate courses, and they are not linked in any way), and some have the lecture linked conceptually to the laboratory portion. Both students and many instructors tend to think of these sessions as separate. Link the two together by using one of the "lecture" periods each week for brainstorming experimental ideas, designing experiments, analyzing data, discussing experimental papers that link the laboratory with the class topics, running tutorials or workshops on things such as presentation of data, writing, and reading the experimental literature. Remember, you aren't trying to "cover" all the content of biology in the classroom; you are helping your students learn the fundamentals of biology as well as how to *do* science. If the laboratory section is an entirely separate course, try to arrange the time within the laboratory section so that there is a short period of time reserved for things that are more conceptual. If the instructors of the laboratory sections are different from those who teach the lecture and you want to link the two conceptually, you can plan the order of topics together. If the "lecture" instructors know what the "laboratory" instructors are teaching, it is possible to coordinate around themes and fundamental concepts. At my institution, our introductory laboratory course instructors have weekly meetings during the semester to coordinate concepts and to share pedagogical ideas. Although these interactions make for a larger time commitment, the benefit of sharing ideas with colleagues energizes the course experience for the students and infuses the course with new ideas and approaches. Instructors can also share the workload in terms of developing new laboratory modules (see Chapter 5).

Time of Academic Year

If you have a field-based laboratory as part of a discovery-based module, the time of year can be a considerable constraint. Going outside to a field site in the fall in a temperate zone is a far different experience than trying to go outside in the middle of February. Frozen soil and water sources cannot be sampled. If your field experience session relies on observing leaves or flowers or even most invertebrates, the time of year for these experiences has to be taken into consideration. As a consequence, the laboratory that is taught in the fall term will not be the same as that taught in the winter term if your institution is at more northern latitudes. At our institution, we have rearranged the order of the laboratory modules to accommodate seasonal restrictions on field-based experiences. On the other hand, seasonal changes could be incorporated into the student projects as experimental variables. For example, students might be curious about the distribution of micro-invertebrates in frozen soil or frozen lake water.

Alternatively, if field-based experiences cannot be performed at your institution because of other constraints (such as transportation to field sites or lack of access to good field sites), you could consider your own institutional grounds as your "laboratory." For example, as described in the following chapter, our institution has areas of lawn that are just above an underground heat delivery system and areas that are not. We measured plant growth along with multiweek measures of soil temperature, snow cover, and sunlight exposure of different plots on campus using inexpensive data loggers. Students analyzed the enzymatic production of cyanide in clover occurring in the different clover populations at these sites. The presence of cyanide in the clover is a known genetic characteristic that correlates with different environmental conditions (see Chapter 5).

The academic calendar limits how many weeks are available for student-designed experimentation. For example, in the fall term at many institutions, there is a break for Thanksgiving that makes it impossible to conduct laboratories for that full week. Multisection laboratory classes that need to be centrally organized because of time and resource constraints would not be able to hold laboratory that week. In addition, partial weeks at the beginning or ends of the semester are not useful for laboratory sessions. Thus, even if an academic calendar is 13–15 weeks from the classroom session point of view, the number of weeks available for laboratory sessions could be as few as 10. The winter/spring semester might have 11 or even 12 sessions, making the spring term longer than the fall term.

The Physical Arrangement of the Teaching Laboratory

The interior design of many teaching laboratories consists of long rows of standing-height laboratory benches (Figure 4.1). This spatial arrangement works well for pairs of student collaborators, but groups of four can become difficult because of the spatial arrangement of the laboratory benches. If your teaching laboratory has a small classroom or even a seated alcove somewhere nearby, you can have small student groups gather in those spaces for planning their experiments, have them divide their work to pairs that work near each other at the laboratory bench.

More modern biology facilities have a different spatial arrangement consisting of circular or modular groupings, such as diagrammed in Figure 4.2. This kind of arrangement works much better for biology and life science laboratory modules and student teams.

The smaller tables can seat four or five students, there is a central area with a small sink and electrical outlets, and often there is under-table storage for equipment such as microscopes. When students are grouped in this spatial arrangement, it's much easier for them to interact as collaborative teams. By having the students seated at desk height rather than at standing height, their line of sight to you is better and you can more easily circulate among groups.

Groups should be no larger than four students, though. Larger than that and the group dynamic does not work as well. One or more students will not be occupied enough and so will disengage from the rest of the group. Also, odd-numbered groups, such as three or five, do not appear to work as smoothly. Four to a group is optimal because they can subdivide into pairs. If your introductory biology laboratories are usually taught in a

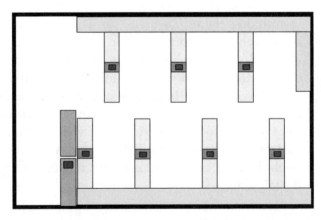

Figure 4.1 Physical arrangement of many introductory science laboratories.

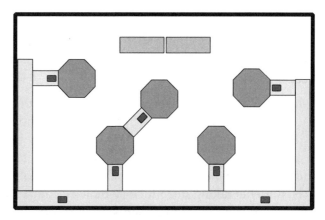

Figure 4.2   Floor map of group-focused laboratory space.

large, standing-bench-style room, you might consider alternative spaces at your institution with a better spatial configuration. Of course, moving your medium to large-sized classes to a different location will have its own downsides: scheduling the space if it is used for other courses, preparation space, storage of supplies, equipment, and ease of access. This may be more trouble than it's worth.

Class Size

Many introductory biology courses meet for lecture sessions in a large lecture hall. Several hundred students might all gather together. These students are then grouped into smaller laboratory sections that might be 30–32 students or so. Some larger institutions typically have enrollments of 500 students or more each semester and so have multiple lecture sections and laboratory sections taught by different instructors. Indeed, many institutions have laboratory sections taught by graduate student teaching assistants who themselves do not attend the lecture class periods. When institutions have multiple laboratory sessions, upwards of 20 or more, making changes in how the laboratory is taught is a huge challenge. The department needs to be committed to making the changes for the entirety of the sections. A number of larger institutions, such as Northern Arizona University, have made such large-scale changes. It required the input and creativity of multiple faculty and staff members over the course of more than 1 year. Faculty received a summer stipend for the planning of the reorganization. Changing the order of laboratory sessions to make them more conceptually cohesive, such as outlined in Chapter 2, does not require such large-scale investments. The payoff in terms of faculty satisfaction and student learning is well worth the initial investment.

Number of Laboratory Sections

In order to offer a hands-on laboratory with an introductory biology course, most institutions offer multiple sections, each with 25–35 students. If your department offers introductory biology with a laboratory, chances are you need to find space and time to teach 120–700 or so students each semester. Even small liberal arts colleges have large numbers of students taking introductory biology courses to fulfill medical school or general college requirements. If your introductory course services 300 students, then you will need ten 3–4-hour time periods scattered through the academic schedule if each laboratory section has 30–32 students. If you have only one laboratory room or do not have adequate equipment or resources to offer two simultaneously scheduled laboratory sections, you quickly run into time constraints in the academic schedule. At my institution, we offer five sections of about 20 students each of our introductory biology laboratory course. We have only one laboratory room that is devoted to the introductory course, and each laboratory period is 4 hours long, so we offer a section every afternoon, with the classroom portions offered in the mornings. Because our laboratory includes field experiences, we cannot offer evening sections. Even if we used our greenhouse, the photoperiod requirements of plants would constrain our teaching laboratories both during the day and at night. I suppose we could have night-shift plants, grown in light-controlled incubators, but there are limits to what's reasonable! Larger institutions or larger biology programs will often have more than one laboratory room, and so they can provide concurrent sections. The demands placed on the rooms, in terms of preparing for the laboratory session, cleaning up afterward, and holding the laboratory session, means that there aren't a lot of times when the rooms aren't occupied. This restricts the independently designed student discovery experiments to the period in which the laboratory is in session.

Resources for Discovery-Based Laboratories

There are never enough resources. Budgets are limited; materials are expensive. Discovery-based laboratories can be cost-effective, though, because having students work in groups conserves resources. The reality in most life sciences laboratories is that science is collaborative. Scientists work in teams. Learning to work collaboratively and in teams is an important skill. The lone scientist working in an empty laboratory in the middle of the night is just a myth.

A typical introductory laboratory is designed for 18–32 students. The most effective student team size is four. Larger groups tend to be ineffective. Eight groups of four requires fewer resources than the actual number of individuals involved, so groups of students are more cost-effective than pairs or individual students. Another way to save on resources is to pool data or samples when using expensive equipment. For example, if you are having your student groups do a PCR assay, combine groups to share a single micro-titer plate.

Let's take an example, developed in Chapter 3: Using *Elodea* as a model system to explore the effects of climate change on plant productivity. Remember we had small teams of students developing experiments to test the effects of climate-related conditions such as acidic pH, higher concentrations of dissolved CO_2, or higher temperature on photosynthesis, chlorophyll levels, and chloroplast number. You can also have two groups collaborate on the discovery-based experiments. If more than one group wants to look at the effects of increasing acidity on *Elodea* photosynthesis, suggest the two groups might combine their projects so that the two projects can be comparable. Maybe one group could test four pHs and the other a different four. If the groups collaborate on experimental design and combine their ideas into one feasible idea, they can cover a wider range of possibilities. This will make research presentations more exciting for the groups and will lead to more lively interactions among groups.

Another way to conserve resources (both supplies and time) is to limit the types of questions that student groups can develop experiments to address. Giving the students a list of possible environmental conditions to explore is particularly helpful for introductory students, who have little experience developing feasible questions to explore. The list also gives you the chance to have materials available in advance. Once you have had a few iterations of the new experimental approaches, you will begin to build a collection of experiments that subsequent groups can build upon (see Chapter 6 for an example of how this works in an intermediate-level neuroscience course). Then, you can provide your student groups with brief summaries of previous work done by student teams, and they can continue the project and build upon the knowledge gained.

Organisms

Not all organisms work well in discovery-based laboratories, particularly at the introductory level. Organisms need to be easy and cheap to grow and maintain. Organisms that don't require high magnification to visualize if you plan to use light microscopy and that grow rapidly or have short life cycles are optimal. Some organisms that I believe are particularly well suited for discovery-based student experiments are as follows:

(1) *Elodea*, duckweed, or other small aquatic plant

These can be maintained in freshwater aquaria right in the laboratory room. Duckweed is also a very common lake organism that students (or you) can collect from the field and characterize. By maintaining aquaria that include aquatic plants, micro-invertebrates, and small aquatic organisms such as snails and fish, for your introductory laboratories, you can create elements of functioning ecosystems for students to study and evaluate.

(2) Other green plants that can be obtained readily such as spinach, basil, onion family plants.

These can either be purchased and brought into the laboratory or the full plants can be maintained in pots. If your institution has a greenhouse, students can plant and maintain these quite easily. Wisconsin Fast Plants, *Brassica rapa* variants, are also widely used green plants that are very useful for student independent experiments on environmental factors that might influence growth and primary productivity. They can also be used to explore genetics and evolution. These plants are marketed for all education levels, so some of your students will likely have encountered them before. *Brassica* subspecies include broccoli, cauliflower, Brussels sprouts, and a host of other agriculturally important varieties. A great laboratory module using *Brassica* is available at this website: http://serc.carleton.edu/genomics/units/cauliflower.html

(3) *Arabidopsis thaliana*

Another fabulous model organism for exploring plant development and cellular structure and genetics is *Arabidopsis thaliana*. With its genome fully sequenced, *Arabidopsis* is a powerful genetic model organism used extensively in research. It is a member of the *Brassica* family, and there are many genetic mutants available for study.

(4) *Microorganisms such as Escherichia coli (nonpathogenic strains) and yeast*

Bacteria and yeast have been mainstays of college laboratories for decades. Easy to culture with very rapid growth rates, they are particularly useful for student-designed experiments measuring growth in response to different environmental conditions. Yeast is useful for studies of respiration and fermentation. Yeast can be examined with a light microscope or could be viewed using available instructional videos (see Further Reading).

Students can observe yeast budding or can stain the yeast to visualize particular cellular components. In addition, there are a large number of fluorescent strains of yeast available. It is also easy to stain nuclei with DAPI. Nuclei can be observed and counted as a measure of cell density or budding rate using a fluorescent microscope.

(5) *Caenorhabditis elegans*

A microscopic invertebrate, this nematode is cheap and easy to maintain and can perform many readily measurable behaviors. Behavioral measurements couple effectively with genetic and even microscopic level analyses. Chapters 5 and 6 describe three laboratory modules using this organism, one for introductory students, and two for intermediate-level neuroscience or neurobiology students.

(6) *Drosophila melanogaster*

Fruit flies are easy to maintain in the laboratory and are the organisms of choice for courses in genetics. Many straightforward teaching laboratories have been developed using this organism. It is also possible to experimentally study selection pressures and evolutionary change with this organism. Genetic mutants are readily available with observable phenotypes such as eye color, wing shape, and bristle morphology, making it easy for students to perform genetic crosses or to evaluate the result of selection on gene frequency. For example, ebony flies have black eyes and are at a competitive disadvantage under normal rearing conditions when combined with wild-type flies. Over the course of a single semester, students can evaluate the frequency of ebony as compared to wild-type flies and can also compute the relative fitness of the ebony flies (see Further Reading). Some student groups could rear these flies under different environmental conditions and examine the effects on gene frequency over time. Shorter laboratory experiences could evaluate genetic inheritance of traits by exploring phenotypic frequencies and gene frequencies using PCR.

(7) Humans (DNA analysis, exercise or physiology, sensory perception)

Students are very interested in themselves, so experiments involving humans are likely to be well received. Most institutions require you to obtain specific approval from the institution's Institutional Review Board (IRB). For DNA-level analyses, you do not want to examine disease-related genes because of privacy-related and bioethics concerns. There are many different genes, physiological processes, sensory, and other behaviors that can be readily explored in an experimental setting, and, of course, a vast scientific and biomedical literature to which to relate their findings.

Equipment

Having the right and enough equipment to hold multiple laboratory sections of 20–35 students is a major constraint for many institutions. Biology experiments are expensive, requiring diverse instruments from PCR machines to dissection and fluorescent microscopes to spectrophotometers to centrifuges. Discovery-based laboratory projects are not much more equipment-heavy than the more common cookbook-style laboratories. The biggest difference is that your groups may be working with different equipment at the same time. For example, one group might be pursuing a predominantly microscope-based project, while another might be performing both behavioral and microscopic tests. It might be trickier to coordinate the use of expensive equipment both within your laboratory and between other courses. If your institution has only a few fluorescent microscopes, for example, you could set up short mini-sessions by dividing your laboratory sections into smaller groupings that come in to learn or use the microscopes. For example, at my institution, we have a room with six fluorescent microscopes. For my intermediate-level neuroscience course, I divide the class in half, with each half coming for a 2-hour time slot during the 4-hour laboratory period. Students then work in pairs or groups of three at the microscopes.

Safety Considerations for Independent Projects

This is a critical concern and often a substantial constraint on independent projects. Some institutions have strict regulations about students working in laboratories unsupervised or after hours, which confines their experimental work to the class times or to times when you can supervise their work. In addition, there are often strict safety requirements for working with chemicals, sharps, glass, bunsen burners, and other potentially harmful agents. Depending on your institution's safety policies, the choice of equipment, the kinds of solutions, might need to be adjusted.

Transportation for Field-Based Studies

These outings can greatly expand the exposure of your students to different organisms. If you plan to have your students conduct field-studies, you might be faced with needing transportation to a field site. The time to travel is an important constraint, as is the number of seats in your van or bus. Often, depending on the class size, you will need more than one driver, which might limit how many field trips can be conducted because of staff availability and time. You also need to factor in transport time to and from the site into the laboratory period available because students might well have another class to get to. If your institution has a greenhouse, you can also schedule "field trips" to have your students conduct various comparative anatomical studies or to set up growth experiments in pots. Growing plants will require advanced planning, possibly having your students set up their experiments early in the semester and revisit their samples periodically throughout the term. If you have access to growth chambers that can control photoperiod, temperature, and humidity, a rich array of possible experiments can be performed. With an inexpensive grow lamp, you can also set up experiments right in the laboratory. You can even use the grounds of your campus as a "laboratory" and have students explore relevant topics such as anthropogenic effects on plant growth, primary productivity, soil organism diversity, and the like by harvesting samples from your campus and analyzing them in the laboratory.

Preparatory Staff

If your institution has preparatory staff to help you with laboratory set up, you are fortunate! Making solutions, placing out equipment and supplies, is time-consuming, particularly if there are multiple laboratory sections during the week. It is far easier for the preparatory staff to just prepare the same old laboratories, so helping them transition to the discovery-based modules will take effort and coaxing on your part. Discovery-based independent projects are much more difficult to prepare for. There could be as many as four or six separate set ups – one for each student group. Each group might require different solutions, different types and amounts of expendable supplies, and even different equipment. You can certainly provide guidance and constraints on what experiments are feasible, but even with that oversight, the set up of independent experiments is far more complex and requires a good deal of flexibility. Because the experimental design is itself being learned, the experimenters are likely to realize that there are flaws with their plans, and they may want to suddenly have different materials to revise their experiments. Your preparatory staff are crucial partners with you; they need to understand what your curricular goals are, what the experiments are that the students have planned, and likely problems that will be encountered. The first time you implement discovery-based experiments usually has everyone scurrying around to modify imperfect plans. If you include your preparatory staff on this process from the beginning, the transition to this kind of laboratory format will be smoother.

Student Interns/TAs

Undergraduate interns or graduate teaching assistants can be an enormous help in introductory laboratories. They can circulate among the student groups, helping with equipment operation, answering questions about procedures, and the like. In order for your interns to be most helpful, they should have a pretty good understanding of the laboratory, its goals, and the techniques. In addition, they should have experience designing and carrying out experiments. You may need to take additional time to train your interns and TAs so that they can be most helpful to you. Depending on their work schedules, the interns and TAs help with laboratory setup and clean up or hold office hours to help students with their experimental designs or data analysis. They can also help you supervise laboratory groups at times when you cannot be present.

Summary

With careful planning and awareness of the constraints, however, it is possible to create a multilevel discovery-based laboratory experience for your students that is authentic and that positions them to learn fundamental concepts and the scientific process much more deeply. The following chapter gives a specific example.

Further Reading

1. Howard, D.R. and Miskowski, J.A. (2005) Using a module-based laboratory to incorporate inquiry into a large cell biology course. CBE 4: 249–260.
2. Instructional video about yeast preparation for laboratory studies: http://www.youtube.com/watch?v=ZrZVbFg9NE8
3. Salata, M. 2002. Evolution lab with *Drosophila*. Bioscene 28(2) 1–6.
4. Micklos, D. Genome Science: A Practical and Conceptual Introduction to Molecular Genetic Analysis in Eukaryotes. Cold Spring, Harbor Laboratory Press.
5. Relyea, R.A. (2005) The impact of insecticides and herbicides on the biodiversity and productivity of aquatic communities. Ecol. Appl. 15: 618–627.

6. Freedman, B. and Lacoul, P. (2006) Environmental influences on aquatic plants in freshwater systems. Environ. Rev. 14.2: 89.

7. Instructional videos on the use of model organisms like *C. elegans, Drosophila* and others. Journal of Visualized Experiments (JoVE): [instructional video: http://www.youtube.com/watch?v=ZrZVbFg9NE8]

8. This site, hosted by Carleton College, has lots of great information and resources for teaching science: serc.carleton.edu/index.html

5 A Model Introductory Biology Course

Several years ago, the Biology department at Vassar College implemented a laboratory course for our introductory students that is modular and focused around current areas of biological investigation. "Introduction to Biological Investigation" is a laboratory-centered course that teaches fundamental concepts in biology, including evolution, structure–function relationships, biological information storage and transmission, and broader systems. We formed a committee to develop the course, and some of us published the course curriculum in 2009 (see Further Reading). This chapter is adapted from that article and includes several additional laboratory modules that we have implemented in subsequent years. In addition, this chapter considers the benefits and the downsides to the course structure, based on formal and informal assessments.

Four to six faculty teach "Investigations" each academic year. Most often, we each teach one laboratory section within a term, but sometimes we might teach two separate laboratory sections, each of about 20 students. Each section consists of a weekly 4-hour laboratory period and a separate 75-min discussion or classroom period. Typically, the department offers enough sections to accommodate 90–120 students each term.

"Investigations" comprises three multilevel laboratory modules – each 4 or 5 weeks in length. Each module is focused around a relevant and current research question based on the published experimental literature in that subfield of biology. Each centers around student-designed experiments guided by the published scientific literature.

Instructor Group Meetings

Weekly instructor group meetings maintain the coherence of the different individual sections of this course. We learn from each other and discuss

Discovery-Based Learning in the Life Sciences, First Edition. Kathleen M. Susman.
© 2015 John Wiley & Sons, Inc. Published 2015 by John Wiley & Sons, Inc.

and trouble-shoot issues that arise over the course of the term. The meetings also provide opportunities to consider new directions we might take these laboratory modules, as well as to evaluate how the student-designed experiments work (or don't work) within the structure of the course. Our shared experiences rapidly create a repertoire of successful student experiments to use as a storehouse of helpful information for future groups of students and faculty. We discuss kinds of course assessments and how well we think the course is working to teach our students the fundamentals of biology in a discovery-based organization. Although the weekly group meetings add to the time commitment each of us invests in the course, they are immensely valuable curricular and professional learning sessions.

Shared Course Materials

Because the laboratory modules represent different fields of inquiry and levels of biological organization, none of us instructors is an expert in all the areas covered by the course. Newcomers to teaching the course find that they are learning and teaching outside their areas of specialization, outside their comfort zone. For example, when I first began teaching the course, only one of the three modules was within my area of scientific expertise. The other two were areas of biology I had never explored beyond delivering lectures or running technique-focused introductory laboratories. Another critical component to the success of this course, then, is the shared instructor materials. The faculty who develop a laboratory module prepare slide sets and instructor reading materials, assignments, experimental articles, student experiment ideas, and other valuable teaching resources. The module developers teach and train the other instructors through the weekly instructor meetings. Faculty instructors usually cycle into the course and teach it several years in a row in order to become fully trained and skilled in the different laboratory modules. Many of us have remarked how much deeper our biological training has become as a result of participating in this course. I certainly feel like a much more knowledgeable biologist, as opposed to the more narrowly defined neurobiologist that I was trained to be.

Flexible Design Allows for the Introduction of New Modules

Each laboratory module poses a relevant and current question in biology and addresses it from more than one level of biological organization. In addition, each introduces students to more than one major technological approach to biology and provides students with training in quantitative analysis and scientific communication. The overall course encompasses most, if not all, of the levels of biological organization and introduces students to many of the major types of techniques that are covered by standard introductory biology laboratory courses. Importantly, students apply the techniques to experiments of their own design. The self-contained modular design makes it easy to add and subtract modules from the overall course while still accomplishing the curricular goals of the course. The flexibility offered by this modular organization allows for the introduction of new areas of exploration and allows the course to keep up with advances in biological investigation. The flexibility enables new faculty members to be able to introduce modules focusing on their areas of biological specialization and keeps the course "fresh" for long-time instructors. Over the course of the 12 years that we have taught "Investigations," we have replaced two of the modules with two newer ones. We now have a total of six different laboratory modules that can be combined in different ways for a fresh and exciting overall course structure.

Overall Conceptual Organization

"Investigations" achieves a number of curricular objectives:

(1) Approach fundamental concepts in biology using an integrative, topic-centered approach, with a strong focus on the relationship between genes and function, and the evolution and inheritance of phenotypic traits. Each laboratory module approaches these concepts from more than one level of biological organization. Across the modules, all major levels of biological organization are included. So, if one module focuses predominantly on cellular or behavioral levels of organization, then another might focus on population or ecosystem levels.

(2) Construction of a "tool kit" for biology students, consisting of skills and concepts that are essential for every biologist and for every informed and educated student, including a thorough training in experimental design, hypothesis testing, data collection and analysis, and scientific communication and writing.

(3) Infusion of an enthusiasm for biology by exposing and involving students in current questions in biology and providing them with the latitude and opportunity to design and carry out experiments that

encourage both self-discovery and real discovery of new biological information.

(4) An emphasis on the *process* of doing science rather than on specific terms and content while learning sufficient content to be conversant in biology and have a firm understanding of fundamental concepts of biology.

Each module begins with a classroom period that introduces the experimental question of focus and the background knowledge required to engage that question. The initial laboratory session introduces the students to the measurement techniques they will use in their independently designed experiments. Then, students read a primary research article centered around the experimental question, discuss that article, and design their own experiments. Figure 5.1 illustrates the overall conceptual organization of each laboratory module.

Each module is self-contained and so can be performed in any order through the semester. This allows us to work around limitations such as time of year for field-based sampling. Each module engages more than one level of biological organization, trains students in the use of one or more current technological or experimental approaches used in biological research, refers to one or more primary experimental articles as guidance and background for student-designed investigations, allows for student-designed experiments using those techniques and approaches, and provides students with quantitative analysis and written/oral science communication practice.

Laboratory Modules for the First Edition of "Introduction to Biological Investigation"

Module 1: *Caenorhabditis elegans*: From Genes to Behavior

- How do organisms move within their environment?
- How do muscle cells and nerve cells coordinate that movement?
- How do genetic mutations disrupt (or change) movement?
- How can scientists study the relationships between genes and behavior?

This laboratory module focuses on structure–function relationships and on information storage and transmission at the levels of macromolecules, cells, and behavior. Students work with one of the most powerful model

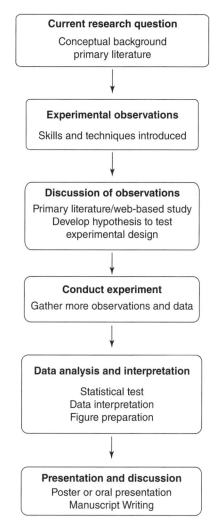

Figure 5.1   Conceptual framework for "Investigations in Biology." (From Ronsheim et al. (2009). Adapted with permission from The Education Resources Information Center (ERIC).)

organisms currently in use by biologists, the nematode *Caenorhabditis elegans*. There are many genetic mutants with phenotypic changes in form and behavior that can be studied.

Many genes are involved in the development, architecture, and regulation of neurons and muscles that allow for different types of locomotory behaviors in all animals. Students learn about how mutations in these genes can result in physiological, anatomical, and functional impairments. They learn a straightforward and quantifiable behavioral assay to assess locomotory function. In addition, they learn how to extract DNA

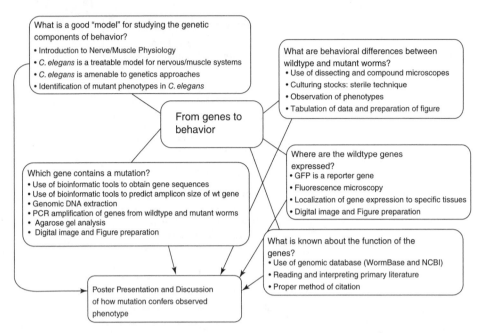

Figure 5.2   Conceptual organization of "From Genes to Behavior." (From Ronsheim et al. (2009). Adapted with permission from The Education Resources Information Center (ERIC).)

from single nematodes and how to use PCR techniques and DNA agarose electrophoresis to identify mutations in key genes involved in locomotion. Students use bioinformatics tools and databases to identify and explore mutations in genes. Finally, the students examine different cells types, namely neurons and muscles, using fluorescent microscopy and digital imaging of green fluorescent protein (GFP)-expressing transgenic nematodes. Then, small groups of students design an experimental approach to compare the behavior of wild-type and mutant nematodes and how to couple that behavioral experiment with a PCR-based examination of what type of mutation results in the behavioral impairments. Because this is the first module of the course, the scope of the independent projects is more limited, and the development of the experimental design is more guided. Figure 5.2, from our published article, illustrates the organization of this laboratory module.

Students learn about model organisms and their importance for studying fundamental biological processes. They learn how to manipulate the nematodes using dissecting microscopes and how to conduct a simple behavioral assay that measures locomotion. Because *C. elegans* have only 302 neurons, most behaviors are governed by only a few cells. When a behavior is disrupted by a genetic mutation, the links between the gene, the cells, and the behavior are readily made. Behaviors such as crawling, swimming, avoidance of touch, chemotaxis, or feeding can be quantified using a dissecting microscope. Students design a simple experiment using one of a selection of behaviors to examine if and how a genetic mutation affects it. The experimental designs include sample size, controls, reliability, and the kind of statistical test they would use to assess differences in the mutant nematodes. Also, in this first laboratory session, the student groups extract DNA from individual worms of each strain, mutant and wild-type, and prepare a PCR reaction to identify specific mutations. Students learn to work with small volumes of liquids using micropippetors, microfuges, and microfuge tubes. The PCR reaction tubes are gathered into the same PCR machine, and one of us places the complete reactions in the freezer until the next laboratory session.

The following session, the students separate their DNA fragments using agarose gel electrophoresis and compare them to a series of DNA size standards. During the time that the electrophoresis process is running, the students work in their laboratory groups solving problems that reinforce their understanding of how polymerase chain reaction works and how it is similar to and different from DNA replication. Students stain and analyze their gels and construct a standard curve from DNA standards for measuring the size of their DNA fragments. We found that the self-discovery process of confirming a kind of mutation (deletion or insertion) was the most beneficial aspect of the electrophoresis identification of PCR product. It was valuable to discuss the effects of large insertions and deletions in particular genes on the function of the resulting protein. This discussion ties in well with a consideration of a primary experimental paper about their mutation and with the following week's laboratory session, which involves an introduction to and practice with the major bioinformatics resources at the National Center for Bioinformatics (NCBI) and WormBase (see Further Reading). The bioinformatics exercises can also be done in the classroom session or as a homework assignment if there is not adequate time to devote an entire laboratory session. Students use some of the powerful predictive tools available at these sites to predict the sizes of PCR products on the basis of the primer sequences to confirm the results they generated themselves. They identify genes and locations within genes where the mutations they are analyzing reside. From this information, they develop ideas about how those mutations disrupt protein structure and function and how that disruption affects cell, tissue, and organism activity.

Our laboratory module also includes a week where students observe the cell- and tissue-specific expression of their gene of interest using transgenic nematodes expressing GFP under the control of cell- and tissue-specific promoters for their gene. So, for example, for the neuron-specific

Table 5.1   **Examples of** *Caenorhabditis elegans* **Mutants with Notable Behavioral Defects**

Gene (allele)	Chromosome	Strain
unc-119(e2498)	III	CB4845
unc-54(e190)	I	CB190
eat-4(ky5)	III	MT608

gene *unc-119*, students examine neurons using an *unc-119::GFP* strain of nematode. Students gain additional practice with microscopy, in this case fluorescent microscopy, and learn how to capture digital images and create well-labeled, high-quality figures. They refer to digital resources such as WormAtlas to identify particular cells that express the GFP. At the conclusion of this laboratory module, students present the results of their behavioral, molecular, and microscopic analyses in the form of a poster. They learn how to prepare figures and figure legends, how to organize their written information in a concise, scientific format, and how to orally present their biological results. The combination of behavioral, microscopic, and bioinformatics approaches provides a rich, integrative, and multilevel way to explore fundamental concepts such as structure–function and information storage and transfer.

This laboratory module engages students in powerful approaches to identifying genes, mutations and the functions they subserve. Although not hypothesis-driven in the traditional sense of the word, the experimental approaches and self-discovery illustrated by this laboratory module expose students to the powerful advantages of model organisms and bioinformatics approaches to studying structure–function relationships. A large part of gene structure–function science involves investigating genetic mutations for functional defects and for structural change to DNA sequence. This module mimics those approaches. Different types of genes and behaviors can be studied by students, either as a whole laboratory group or in individual groups. If you have additional time to devote to this laboratory module, student groups can select from a list of mutant strains that have various genetic defects (Table 5.1). They learn about the gene's function, select primers (or design them) to identify the mutation in their mutant strain of nematode, and then design and carry out a behavioral assay to examine the mutation's effect on behavior.

Summary
Students learn about DNA structure and the consequences of mutations, how mutations arise during DNA replication, how gene expression is regulated in different tissues, relationships between protein structure and

function, and how mutations interfere with protein function. They also learn about how neurons regulate muscles and how muscle contraction occurs at the level of the protein components. In addition to the fundamental concepts and experimental experience, the students also gain important practice presenting data and information in written and oral forms.

Module 2: Cyanogenic Clover: Genetic Variation and Natural Selection

- How do plants defend themselves against voracious herbivores?
- How might these defenses have evolved?
- What factors in the environment or in a population of plants influence the expression of plant defenses?

Plant defense mechanisms, their constraints and advantages from an evolutionary point of view, are active areas of current ecological research. This second laboratory module examines the environmental and genetic influences on the cyanogenic plant defense system found in white clover (*Trifolium repens*). We adapted this laboratory module for "Investigations" from approaches originally conceived of by Kakes (1991) (see Further Reading). In white clover, plant cells compartmentalize a nontoxic cyanogenic glucoside, linamarin (A), away from the enzyme linamarase (L), a galactosidase sequestered in cell walls. When cells are damaged by tearing or crushing, the linamarase is released from the cell walls and catalyzes the breakdown of linamarin (A), producing acetone cyanohydrin, which spontaneously breaks down to acetone and cyanide gas, an effective plant defense against small herbivores. The cyanogenesis enzyme system is determined by two loci in a Mendelian inheritance pattern. Plants with at least one dominant allele at each locus are cyanogenic. The presence of active cyanogenesis in clover varies with altitude and geography. Students read the published literature describing altitudinal and latitudinal variation in cyanogenesis among clover strains in Europe and North America (see Further Reading). On the basis of discussion of those articles, student teams brainstorm possible explanations for the variation. Does the winter soil temperature influence the frequency of cyanogenesis? Might genetic drift or gene flow influence the frequency of cyanogenic clover in a region? Does herbivory play a role? By examining the presence of cyanogenesis in clover leaves sampled from different sites on the campus, or from different cultivars obtained from an international seed bank, students can explore questions of natural selection (or mutation, gene flow, drift) and environmental conditions (such as presence of

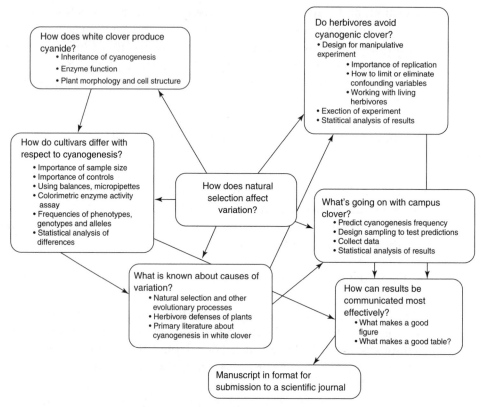

Figure 5.3   Conceptual organization of "Cyanogenic Clover". (From Ronsheim et al. (2009). Adapted with permission from The Education Resources Information Center (ERIC).)

herbivores and effects of temperature) that might influence the presence or absence of cyanogenesis, assayed by a straightforward colorimetric enzyme assay. Figure 5.3, from the article we published, illustrates the pedagogical structure of this laboratory module.

After the brainstorming sessions, small groups of students design and carry out independent experiments exploring these questions. In one set of experiments, students obtain temperature data using inexpensive long-term data loggers we buried at different sites on campus with clover populations. In another set of experiments, students explore whether a natural herbivore of clover, a locally abundant snail (*Cepaea nemoralis*) or a commercially available beetle (*Zophobas morio*), avoids cyanogenic clover in a taste preference assay from clover samples they gather at different sites on campus. Students could test additional environmental conditions including soil water content or degree of sunlight exposure and the frequency of cyanogenesis. They could also examine the production of cyanide at different life stages of the plant (i.e., young plants vs older

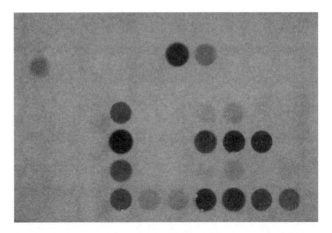

Figure 5.4   Example Student Cyanogenesis Assay Results. The circular spots indicate production of hydrogen cyanide gas. Intensity of the spot indicates genotype, with darkest spots reflecting homozygosity for both alleles, lighter spots reflecting heterozygosity of one or both alleles. (Courtesy: Dr. David Jemiolo.)

plants), or in response to injury over time, or they could examine the distribution of the enzyme system in different plant tissues. If your campus is not amenable to this sort of field-based sampling, it is possible to obtain different cultivars from the USDA and grow plants in your campus greenhouse or in growth chambers.

Kakes (1991) developed a straightforward colorimetric assay using Feigl-Anger test papers for cyanide gas production that has been adapted more recently as a high throughput technique to screen agricultural and natural variants for cyanogenesis (Olsen et al., 2007, Further Reading). Filter papers are soaked in a solution of tetra-base (4,4′-methenebis-(N,N-dimethylaniline) and copper(II) ethylacetoacetate. In the presence of hydrogen cyanide gas (HCN), a colorimetric reaction occurs, resulting in a blue spot on the test paper (Figure 5.4)

The assay, although semiquantitative, is very easy for students to perform to determine phenotype and genotype of different clover populations from samples they collected in the field or that they obtained from cultivated greenhouse varieties. Students measure similar amounts of clover leaves (without the stems) using a microbalance and then place them in a salt-balanced solution in microfuge tubes. The samples are frozen at −80°C for at least 30 min (this period of time can be much longer if needed) and then rapidly thawed and mechanically disrupted with a toothpick or tissue-crusher, releasing the linamarase and linamarin into the solution. The cyanide gas produced interacts with the paper, producing blue dots, the intensity of which reflects the degree of enzymatic activity (Figure 5.4). Students can place several aliquots of

Table 5.2   Genotyping Using the Cyanogenesis Paper Assay

Genotype	Extract Only	Extract + Enzyme	Extract + Substrate
LLAA	Bright blue	Bright blue	Bright blue
LLAa	Blue or bright blue	Blue or bright blue	Brighter blue
LlAA	Blue or bright blue	Brighter blue	Blue or bright blue
LLaa	White	White	Blue or bright blue
llAA	White	Blue or bright blue	White
LlAa	Light blue	Brighter blue	Brighter blue
Llaa	White	White	Light blue or Blue
llAa	White	Light blue or Blue	White
Llaa	White	White	white

these solutions into the wells of microtiter plates. To one aliquot nothing is added, to one external linamarase is added, and to a third external linamarin is added. Groups can place many different samples in the same microtiter plate, representing their treatments or study sites or whatever their experimental variables are. They place a piece of the test paper over the wells, seal, and incubate at 37°C for about 30 min to 2 hours or so. The resulting pattern of dark blue, light blue, or white spots over each well indicates the presence of strong, moderate, or no cyanogenic activity (Figure 5.4). The genotypes can be inferred as indicated in Table 5.2. From the assay, students can determine phenotype and can infer genotype to a large degree, although some of the results might be ambiguous. These results can be compared to other components of their projects, such as the herbivore preference assay, or a study of freezing effects on cyanogenesis, or the effects of age and many other ideas. Students can prepare a poster of their projects and research findings, or they could write up a paper in the style of a peer-reviewed article.

Summary

During this laboratory module, students learn the fundamental concepts of evolution, natural selection, Mendelian inheritance of traits, ecosystem and community-level interactions, enzyme structure and function, cell metabolism and cell structure, and functional compartmentalization. They learn more complex quantitative analyses such as goodness of fit tests, Chi square or t-test, and even ANOVA. They gain more practice in communicating their research findings by writing and preparing figures and a manuscript or poster. In addition to white clover, many other plants are cyanogenic: cabbage, cauliflower, bamboo shoots, cassava, mustard, turnip, radish, and many others. This module is ripe for exploration.

Module 3: Biodiversity and Soil Microbial Ecology

- What environmental factors influence the biodiversity of soils?
- How does soil ecology influence the above-ground biodiversity of an area?
- How do issues of scale and spatial heterogeneity affect biodiversity?

Soil is everywhere. This laboratory module can be performed in any climate, on any campus. The main biological focus of this field-based laboratory module is how local variations in environmental conditions affect the biodiversity of soil organisms. Your campus can be your laboratory. This module (Figure 5.5) begins with a discussion of biomes, ecological communities, and global patterns of biodiversity. We focus on soils and soil organisms as foundational features of any biome. Students read the experimental literature about the relationship between

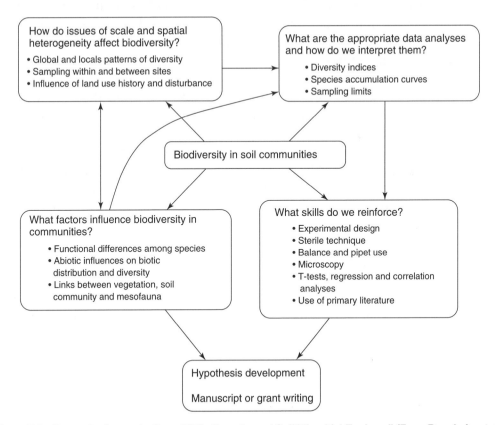

Figure 5.5   Conceptual organization of "Biodiversity and Soil Microbial Ecology." (From Ronsheim et al. (2009). Adapted with permission from The Education Resources Information Center (ERIC).)

soil organism biodiversity and global nutrient cycling (Wardle et al., 2004; see Further Reading). The soil ecosystems determine the supply of soil nutrients that fuel the above-ground ecosystems grounded in plant primary production. In turn, the organic material provided by the above-ground ecosystems determines the health and functioning of the soil ecosystems. The complex, multilevel relationships between below- and above-ground communities establish and maintain global nutrient cycles. How human activities affect this vast interconnectedness is a matter of critical importance. Many students are concerned about how, for example, slash and burn agriculture in the Amazon will affect overall primary productivity or how the increasing levels of atmospheric CO_2 from industrialized nations such as the United States and China will influence global productivity. However, most are not aware of the relationships between above- and below-ground systems and how these relationships change as a result of human activities.

In this laboratory module, students sample soil, measure abiotic characteristics such as pH and water content, and assess the kind of soil for the geographic area and the underlying bedrock using geological survey data for the region. They also cultivate some of the bacterial and fungal communities that are present. We have used plate dilution assays to construct a class morphospecies laboratory, but it is also possible to use a molecular assessment of bacterial diversity, using PCR and sequencing to identify 16S rRNA variants using soil DNA kits followed by PCR and sequence analysis (Fierer and Jackson, 2006; see Further Reading). Because this molecular technique is more expensive and more complex, this approach may be less feasible for introductory laboratories. To pursue this molecular approach would require a budget that would allow you to send the PCR products out for sequencing. The module I outline in the following section assesses microbial diversity using the plate-based morphospecies approach. While less exhaustive a diversity survey, it is highly instructive about types of microbes and works quite well in the teaching laboratory.

After learning these basic techniques, student groups design a study to examine variations in soil pH or water content at different topological sites or different soil types sampled on campus and correlate those measurements with biodiversity of bacterial and/or fungal communities. They learn how sampling techniques, such as a stratified random sampling method, are integrated into the design of a study. They examine the above-ground biodiversity by collecting leaf litter to assess mesofaunal organisms and by assessing tree and other plant diversity in the sampling regions. We make use of soil and topographic maps, as well as historical aerial maps to understand the underlying geology of the area, as well as recent changes in land use.

In more recent iterations of this laboratory module, students have sampled from local sites with different levels of herbivory from white-tailed deer or different levels of human disturbance (e.g., near a road site vs in the woods). There are many possible variables that you can have students explore depending on your own campus environment and on the research questions your students wish to pursue.

We have found it most useful for our student groups to collaborate on assessing biodiversity. So, at the beginning of this laboratory module, before going out into the field, we have a class brainstorming session. What aspects of biodiversity do we want to sample? How can we generate a larger sampling area by working together as a full laboratory section? How will we collect our data to ensure we can combine our individual group data sets?

Lots of data are generated. From Berlese funnels, they identify mesofaunal diversity from the collected leaf litter samples. From the microbial (fungal and bacterial) morphospecies libraries, students evaluate morphospecies richness and diversity, constructing Shannon-Wiener and Jaccard indices. Students reinforce and extend their statistical analytic skills by performing t-tests, or ANOVA, or correlation/regression analysis on the data. The culminating assignment for this module is for students to write up their projects as a manuscript. The assignment builds upon earlier writing assignments and focuses on aspects of data interpretation and presentation of results. This laboratory module, similarly to the others, can be modified to accommodate different interests and expertise. Different soil properties, such as levels of soil arsenic or other pesticides, can be measured. Or, the modules could include an experimental angle, such as culturing soil microbe samples under different plating conditions to test hypotheses about what environmental factors might influence microbial diversity.

Additional Laboratory Modules

Because of the modular design of our "Investigations" course, it is possible to retire some modules and introduce new modules to keep the course fresh and vital. For example, a number of years after we introduced the course into our curriculum, we retired the *C. elegans* genes to behavior module and substituted instead a Human SNP Analysis module (Module 4, in the following section; also see Chapter 3). The module retained the molecular and bioinformatics focus of "From Gene to Behavior" and also retained the emphasis on genes, phenotype, and genotype.

Module 4: Personal Genomics: Understanding Individual Genetic Variation

- What can SNPs tell us about human evolution, migration, and disease?
- How can SNPs lead to disease?
- Why is an understanding of SNPs and genetic variation, along with the ethical implications of this understanding, essential for this century's medical care, health insurance, and even other social institutions (such as the criminal justice system, business, and education)?

This module was adapted from a laboratory sequence originally developed at Carleton College as part of a Teagle Foundation initiative (Banta, L.M. et al; see Further Reading) and incorporated into the "Investigations" course to replace the *C. elegans* Genes and Behavior module. Similar to "From Genes to Behavior," students learn about genetic variation, sources of genetic variation, and the relationship between individual genotype, phenotype, and population dynamics at the genome, macromolecular, and cellular levels.

The simplest types of genetic variation are SNPs, or single nucleotide polymorphisms, which occur at particular positions within the human genome with a frequency of more than 1% of the population. Many SNPs are associated with variations in disease risk, such as Alzheimer's disease, breast cancer, multiple sclerosis, or arthritis. Others of the millions of SNPs in the human genome are associated with food sensitivities or memory abilities and other common variations in human behavior and abilities. The vast majority has no known functional or disease correlates, and these make great candidates for student exploration using polymerase chain reaction, DNA sequencing, and bioinformatic analysis.

This module focuses on a SNP in the non-disease-associated Cdk3 gene. In the first week of the laboratory, students extract DNA from their own cheek cells and perform PCR using gene-specific primers to amplify a small region of the Cdk3 gene containing a SNP. The samples are sent to a company for sequencing. The SNP introduces a difference in the restriction fragments generated in an RFLP analysis, which the students conduct the following week in laboratory with part of their PCR product reaction. Students learn about DNA replication and how it compares with PCR, as well as how restriction enzymes work. They visualize their restriction digests using agarose gel electrophoresis. Students analyze their gels and also perform a bioinformatic analysis on their sequence data for the subsequent two laboratory sessions. They compare the two different techniques and also compute the allele frequency across the different class sections to discern whether evolutionary-level change is represented by the SNP analysis. Through this analysis, students learn about population sampling, genetic variation, and Hardy–Weinberg equilibrium. Because the focus is on human DNA, students can examine disease-related SNPs using this bioinformatics approach, an idea that was described in Chapter 3. In class, you can discuss the connections between DNA variation and human medical conditions. Students working alone or in groups design and carry out a bioinformatic study of a SNP of their choosing. Students learn about the growing number of commercial companies that provide personal genotyping services and can discuss some of the scientific and ethical aspects of personal genomics. The published literature is replete with examples of SNPs linked to human disease that can make this laboratory module a relevant and exciting self-discovery experience for students.

Module 5: Behavioral Variations Within a Species

- How do animals maximize the fitness of their offspring?
- What adaptations might be selected that influence oviposition choice? What advantages are there to flexibility in choice of oviposition substrate? Disadvantages?
- What are the relationships between the structures of oviposition machinery and their function?

This 4-week module was introduced into the "Investigations" curriculum as a replacement for the "Cyanogenic Clover" module and was adapted from a published laboratory sequence developed by Blumer and Beck (see Further Reading). Like the Clover module, this discovery laboratory series examines questions of organism interactions and physiological adaptations to environmental conditions and whether those adaptations are heritable.

Insect pests devastate essential agricultural crops and cause substantial economic loss all over the world. As a result, pesticides are a multibillion-dollar enterprise, and their overuse has contaminated our soil, water, and air. Understanding the behavioral and physiological processes that lead harmful pests to infest particular crops is an important way to reduce their impact economically and to mitigate the need to use insecticides that have unintended effects on beneficial insects, wildlife, and human health.

The most common insect pests are "holometabolous," meaning that while adults are winged and highly mobile (e.g., flies, moths, and beetles), juveniles are wingless and much less mobile (e.g., maggots, caterpillars, and larvae). Because the juveniles of such species do not move much, the behaviors that lead adult female insects to lay their eggs (oviposit) on particular plants or seeds determine the food and shelter their offspring will have access to until they reach adulthood. This means that the preferences of adult female insects for oviposition locations are crucial components of evolutionary fitness. Natural selection will act on insect populations such that female insects' egg-laying decisions optimize the number of eggs they lay and the amount of resources their offspring will receive. In this module, students study the oviposition behavior of bean beetles, *Callosobruchus maculatus*, a globally distributed member of the largest order of Insecta, Coleoptera. Female insects lay eggs on various species of legumes and are highly adapted to particular legume varieties, including *Vigna unguiculata* (black-eyed peas) and mung beans, allowing students to ask questions about physiological adaptation, behavior, and natural selection.

Students learn about the life history of two different varieties of bean beetle and compare reproductive morphology and oviposition behavior. These varieties, one native to Asia and one native to Africa, have been maintained for nearly 30 years on particular varieties of legume, resulting in physiological, developmental, and behavioral adaptations that influence oviposition behavior. Students compare the structure of female ovaries between the bean beetles using a fluorescent phalloidin staining technique that illuminates fine cellular and subcellular architecture. They address questions such as whether there are morphological differences that underlie functional differences in oviposition, larval growth, and egg-hatching dynamics. This comparative analysis of structure and function (behavior) relationships introduces students to fluorescent microscopy as well as dissecting microscope use for dissection and microscope specimen preparation. Students also learn a simple oviposition preference assay and can then design experiments that they conduct in the third week of the module to investigate questions such as whether specific bean beetle populations exhibit flexibility in oviposition preference, whether larval hatching and growth are influenced by the female choice of bean for oviposition, or whether the female can alter the number of eggs deposited on the basis of an assessment of the size of the bean. They could also explore the effects of different environmental conditions such as temperature and humidity on oviposition or choice of bean. Many other experimental questions are possible. Students continue their learning of statistical analysis such as Chi-square tests and correlational analyses and write up their findings in the form of a scientific manuscript or a poster for presentation. This module is organized similarly to the clover cyanogenesis module, in the sense that students learn a straight-forward assay that they can then apply to address a student-driven experimental design. In addition, like the clover cyanogenesis module, students examine the question at a cellular microscopic level as well as at the levels of organism, behavior, and population.

Assessment of Learning of Core Concepts and Skills

We assess learning in a number of different ways. During each laboratory module, students take short quizzes or complete homework assignments. These assignments focus on course content and key biological concepts. Each laboratory module also has a culminating assignment, in the form of a poster presentation or a scientific manuscript that gives students practice in written and oral demonstrations of their understanding of both the specifics of their own experiments and the key biological concepts addressed by their experiments. There is also a final cumulative practical examination that evaluates skills learned as well as key biological concepts. One measure of their mastery, then, is how well they perform on these various class assessments. Other measures include student evaluations, both right at the end of the term that they took the course and also further along in their academic careers. Another key assessment is a faculty assessment of whether the students appear adequately prepared to engage more advanced-level study in the discipline.

Student Evaluation of the Course

"Investigations" is so radically different from what most college students experienced in high school that often students initially express frustration because they are not being "lectured at." Instead, students work with biological concepts and apply those concepts to designing and carrying out experiments. They learn fundamental concepts through applying them to a particular topic and biological question. Particularly at the beginning of the course, students will wonder if they are learning "enough" biology. Most students are used to memorizing lists of biological terms and concepts in the absence of a unifying theme or question. In addition, their conception of a laboratory is to follow a procedure that is different each week and to fill out worksheets with questions to answer and tables for measurements. The open-ended and question-driven nature of the laboratory experience in "Investigations" is highly unusual, and students wonder how much biology they are learning if they are mostly asking questions that do not have ready-made answers to them. Rather than memorizing lists of terms and concepts, students use those terms and concepts as they plan, carry out, analyze, and present their research findings. The learning becomes implicit rather than explicit.

We assessed perceived student learning and mastery of concepts and skills after the course had been in place in our curriculum for a few years (see Further Reading). We found that the majority of students perceived their level of familiarity and mastery of key biological skills such as measurement, statistical analysis of data, experimental design, microscopy, polymerase chain reaction, gel electrophoresis, presentation of data in the form of figures and tables, and writing scientifically to be substantially enhanced as a result of taking " Investigations."

Faculty Concerns and Discomforts

In order to teach this course, the faculty have had to radically change their approach. There is no time to give a series of content-rich, term-laden lectures. For example, when we introduce the concept of DNA replication and PCR, we cannot lecture our students about nucleotide structure or every step in DNA replication or the history of the discovery of DNA. Instead, we talk about how replication and PCR are similar and different in the context of a group activity. Students have to learn or review on their own the underlying macromolecular structure that makes DNA work as an information storage and transfer molecule. As a result, some faculty feel uncomfortable that not every concept or term they taught in previous lecture-heavy biology courses is addressed by the laboratory module-focused course. If the modules do not include lectures on enzyme kinetics, for example, then we feel that there must be a large gap in our students' education. This is a very difficult feeling to shake off. It is absolutely true that in order to make time in a course curriculum for students to learn by doing, for students to do science, to practice science, other aspects of the course material have to go. A number of us report that students who have taken the "Investigations" course are well prepared to embark on future laboratory work, particularly the data analysis aspects, but there are gaps in their content-based knowledge. Students taking biochemistry, for example, have not memorized (and then forgotten) the steps of glycolysis or the Krebs cycle in the introductory course. On the other hand, we have all had the experience that the majority of our students who took a standard lecture-heavy content-focused introductory course does not really understand glycolysis anyway. Most often, the students have forgotten the cycles, having memorized them for examinations they crammed for. In contrast, our students overwhelmingly see the interconnectedness of biology, the relationships of form and function, of genes and inheritance. They have a conceptual framework that facilitates deeper learning in biology and a facility with the process of discovery. It works.

Further Reading

1. Ronsheim, M.L., Pregnall, A.M., Schwarz, J., Schlessman, M.A. and Raley-Susman, K.M. (2009) Teaching outside the can: a new approach to introductory biology. Bioscene 35(1): 12–22.
2. WormBase (http://www.wormbase.org/#01-23-6) and WormAtlas (http://www.wormatlas.org/)- these are wonderful resources for students to learn about the model organism, particular cells (and GFP expression transgenic strains), and particular genes. Students can connect to other databases, including the NCBI sources (http://www.ncbi.nlm.nih.gov/) and PubMed (http://www.ncbi.nlm.nih.gov/pubmed).
3. Kakes, P. (1991) A rapid and sensitive method to detect cyanogenesis using microtitre plates. Biochem. Syst. Ecol. 19: 519–522.
4. Bertoni, G. (2010) Got the blues? a high-throughput screen for cyanogenesis mutants. Plant Cell 22: 1421.
5. Olsen, K.M., Sutherland, B.L. and Small, L.L. (2007) Molecular evolution of the Li/li chemical defense polymorphism in white clover (*Trifolium repens* L.). Mol. Ecol. 16: 4180–4193.
6. Wardle, D.A., Bardgett, R.D., Klironomos, J.N., Setala, H., van der Putten, W.H. and Wall, D.H. (2004) Ecological linkages between aboveground and belowground biota. Science 304: 1629–1633.
7. Fierer, N. and Jackson, R.B. (2006) The diversity and biogeography of soil bacterial communities. PNAS 103: 626–631.
8. Banta, L.M., Crespi, E.J., Nehm, R.H., Schwarz, J.A., Singer, S., Manduca, C.A., Bush, E.C., Collins, E., Constance, C.M., Dean, D., Esteban, D., Fox, S., McDaris, J., Paul, C.A., Quinan, G., Raley-Susman, K.M., Smith, M.L., Wallace, C.S., Withers, G.S. and Caporale, L. (2012) Integrating genomics research throughout the undergraduate curriculum: A Collection of inquiry-based genomics lab modules. CBE-LSE 11: 203–208.
9. Blumer, L. S., and Beck, C.W. (2009) Bean beetles: a model organism for undergraduate laboratories in ecology, evolution, and animal behavior. http://www.beanbeetles.org/ Morehouse College and Emory University, Atlanta, GA. Supported by NSF grant DUE-0535903.

6

Two Model Scenarios for an Intermediate-Level Life Science Course

The intermediate-level life sciences curriculum is usually the place where subdisciplines are considered separately and more deeply. Courses in Animal or Plant Physiology, Biochemistry, Ecology, Developmental Biology, and Genetics are common. These are the bread and butter courses of the department, course descriptions boasting a long list of content and techniques specific to a discipline. Students taking these courses have taken introductory-level courses in Biology, Chemistry, and the like. At many institutions, the discipline-specific intermediate-level courses do not have a linked laboratory. Rather, the laboratory is a separate course, most often taken by students majoring in the discipline or even subdiscipline. The smaller colleges and universities still retain the classroom component linked to the laboratory. The laboratory component or course introduces students to the major techniques and approaches specific to that subdiscipline. Many allow for some practice designing experiments, but quite often, there is not adequate time devoted to these experiences to really enable students to engage the experimental process deeply. Students perform limited experiments, analyze data, and write up the results as a laboratory report in the form of a scientific manuscript, sometimes also as an oral presentation. Students gain valuable exposure to techniques and develop writing and presentation skills, but the experimental sessions are very carefully scaffolded. For example, in an animal physiology course laboratory exercise, students might learn how to measure muscle contraction using a frog gastrocnemius muscle preparation. For an independent experiment, students might select from among several different chemical solutions (such as acetylcholine or a cholinergic agonist or antagonist) and measure changes in muscle contractility. They

Discovery-Based Learning in the Life Sciences, First Edition. Kathleen M. Susman.
© 2015 John Wiley & Sons, Inc. Published 2015 by John Wiley & Sons, Inc.

learn through experimentation how different chemicals influence muscle contractility and so reinforce concepts of muscle physiology. However, do they learn the process of physiological research? Do they refine and further their skills in experimental design? Not really.

This chapter describes a discovery-based approach to teaching an intermediate-level course in neuroscience and behavior. This course design can be adapted for many different life sciences subdisciplines. The emphasis of the laboratory is on the design of experiments rather than on the acquisition of a list of skills and techniques. The student projects focus on a current question of interest in the field of inquiry, and the laboratory resembles how a graduate student would be trained.

Neuroscience and Behavior 201, the entry course to the major with the same name, is taken by sophomores and juniors after they have completed courses in introductory biology, introductory psychology, and a course in physiological psychology. There is no textbook for the course; rather, it is based on an integrative, experimental analysis of the primary literature in topics of neuroscience. One key goal of the course is to integrate across a wide spectrum of levels of biological organization, including evolutionary, ecological, organismal, systems (emphasizing but not limited to neural), behavioral, cellular, and subcellular levels. In reading the experimental literature, students examine hypotheses being tested and the underlying assumptions that influence them, the methods and techniques, and the interpretations of results. They learn to integrate across different studies to arrive at insights and understanding of how nervous systems work and how behaviors come to be. The laboratory component centers around a multi-week independent project related to at least one of the main topics of the course. Most often, the laboratory techniques stem from the instructor's own area of expertise. In many ways, the laboratory component is designed to resemble working and learning in a research laboratory. What follows are two different laboratory sequences that worked very well with this course. The key learning goals for the course are to:

- achieve a solid foundation in the experimental approaches to a variety of current research questions in neuroscience and behavior;
- achieve a sophisticated ability to read and interpret the primary experimental literature in the field;
- formulate a hypothesis, design, and conduct a multilevel experimental project over several weeks to discover new information about the relationship between genes and behavior;
- perform and understand appropriate statistical analysis of behavioral data, gain confidence in the use and limitations of model organisms,

computational and bioinformatics approaches to explore evolutionary relationships between genes and behavior; and

- become facile in the "language" of neuroscience and behavior, with a thorough mastery of our chosen subtopics, as well as a keen ability to speak and write on the discipline.

Model 1: Exploration of Gerontogenes and Behavior

In this version of the course (see Further Reading), a large topical focus in the classroom is on the integrative topic, "Genes, Theories and Mechanisms of Aging." In the classroom, we examine evidence of both the effects of aging on nervous system function and behavior and the role of the nervous system in aging. The evidence spans organisms including humans and ranges from genetic to physiological to organismal to evolutionary. Almost all organisms have a finite lifespan punctuated by phases of growth, development, reproduction, and senescence. Despite this universality, lifespans vary enormously among animals. Why? How is lifespan regulated? Recent work has identified a small number of genes that appear to regulate lifespan, and their expression is found in reproductive and intestinal tissues, and, surprisingly, in the nervous system. Intriguingly, manipulation of these genes' expression in the nervous system can influence lifespan (Wolkow et al., 2002; Taguchi et al., 2005; see Further Reading). Researchers in this field of neuroscience have exploited the model organism *Caenorhabditis elegans* to gain insights into some of the genes and cellular mechanisms involved in aging, as well as the role played by the nervous system. These genes are called, "gerontogenes." Manipulation of these genes in *C. elegans* affects lifespan (Wolkow et al., 2002) and responses to environmental stressors such as heat stress and UV light (Braeckman et al., 2001). The topic of aging is a great scaffold for learning about nervous system and behavior.

C. elegans is the model organism for the multi-week student projects. There are a number of available mutant strains for gerontogenes that are expressed in neurons that we acquired for the laboratory projects. Students form small groups and select a mutant strain to examine for their projects. They access databases such as the NCBI and WormBase to investigate what is known about their candidate gene. On the basis of that information, students develop a multiweek experiment to examine some aspect of behavior and nervous system function.

The first few laboratory sessions provide the students with a "toolkit" of approaches that they then apply toward their independent investigations. During the first week, the students learn about *C. elegans*, observing them using dissecting microscopes, using online databases such as WormBase and WormAtlas to learn a bit about their lifecycle, their nervous system, and their genetics. In addition, they use this time to learn about the various genetic mutants available for a study of gerontogenes in the nervous system. This investigation results in the beginnings of a bibliography of recent experimental articles that serves as a starting point for developing their own research questions.

The classroom supports the laboratory. Students read and discuss primary experimental articles on the relationship between aging and the nervous system in many different organisms, from *C. elegans* to mammals including humans. The articles range from genomic to brain region to behavioral analysis. This direct connection between the classroom and the laboratory deepens the learning of fundamental principles of neuroscience as well as enhances the relevance of the student projects to the broader field.

After the students submit their initial experimental designs for their projects, we consult with each group to discuss and refine their experimental plans. These brainstorming sessions enhance both the level of engagement the students have toward their projects and the quality and feasibility of their experimental plans. The groups, over the next 3 weeks, conduct their experiments. Some groups have explored egg-laying behavior, governed by serotonergic neurons, or the response to heat stress or UV light exposure or anoxia. Because each group designs a unique set of experiments, setting up and providing them with the materials they need require careful planning.

An important emphasis of their projects is the selection of a genetic strain of *C. elegans* with a mutation in a gene expressed in the nervous system that has been implicated in aging. The students conduct a comparative bioinformatics study examining the protein sequence similarity of their gene product across at least 10 different animal taxa. They consider the following questions with their analysis:

(1) Are there regions of the protein sequences with strong similarity or alignment? What are the implications for these regions of the protein in terms of overall protein function across taxa?

(2) Are there regions of the protein sequence that appear quite different across the taxa you examined? What might these differences mean?

(3) What does your protein sequence similarity analysis tell you about changes in the gene/protein across the taxa you explored?

(4) How might this protein sequence analysis inform you about the role of the protein in nervous system function or aging?

The 5-week projects culminate in a research symposium where the student groups present their projects in professional research talks. Lively discussion ensues because each project is different and yet all are related to the overall topic of gerontogenes in the nervous system and all used the same model organism. The results are novel, and over the years, a number of the students have pursued their projects further through independent research opportunities as seniors.

Assessment of Skills and Student Learning

Does this student discovery-based laboratory format really enhance learning and understanding? Well, in a word, yes. Students transformed from hesitant learners tentatively working through a procedure to confident researchers independently planning and carrying out experiments. Student confidence in the techniques and approaches was substantially increased, according to both their own reports to the instructors and their responses on surveys we conducted (Raley-Susman and Gray, 2010, Further Reading).

Model 2: How do Common Lawn Chemicals Affect the Behavior and the Nervous System of C. elegans?

The beauty of the format of this intermediate-level course is that it can readily be adapted to different topics and laboratory techniques. After several iterations of the genes of aging topic, I changed the focus to environmental neurotoxicology and sensory behavior. I made some adjustments to the structure of the laboratory to more explicitly distinguish between the techniques and the scientific exploration process. Modeled more closely on how a graduate student would be trained in a research laboratory, this version of the course introduces the toolkit of techniques as workshops and graduate-school-style journal clubs, and then student groups design and carry out a half-semester research project approximately 7 weeks long (Table 6.1). As before, the classroom portion of the course supports the laboratory projects and provides multiple perspectives to deepen the students' knowledge of their projects while also focusing on fundamental concepts in neuroscience within the context of the topic.

By having the initial skills workshop sessions, students know that they are learning skills that they will then practice and put to use. The goal of the laboratory as a whole is not to expose students to an exhaustive list of techniques. Rather, the goal is to learn some techniques and approaches

Table 6.1   The Laboratory Organization

	Laboratory Topic	Goals
Week 1	Workshop I: Introduction to the model organism *C. elegans* and Sensory Behavior Assay	Learn a behavioral assay Design an experiment to explore the sensory behavior in a mutant strain compared with the wildtype strain of worm Statistical analysis (t-test, one-way ANOVA), figure preparation
Week 2	Workshop II: Sensory Behavior II	Learn another sensory behavior assay and conduct another experiment Deepen the statistical analysis, figure preparation
Week 3	Journal Club of small groups of students	Read and discuss a primary research article related to the upcoming independent projects. Brainstorming project ideas during small group discussions
Week 4	Microscopy Skills Workshop	Learn how to prepare specimens for fluorescent microscopy Close observation of wildtype nervous system using GFP-expressing nematodes for different neuron groups Prepare digital images, make a figure and interpret the figure in writing
Week 5	Experimental Design Planning Sessions	Refinement of experimental design in small-group brainstorming sessions
Weeks 6-9	Students Conduct Independent Projects	Behavioral analyses Microscopic analyses Opportunity to repeat, refine, revise or extend experiments to develop a substantial body of work
Week 10	Writing Workshop	Peer-review Focus on how to present the project in both written and oral form
Week 11	Student Research Symposium	

that can be used to address a scientific question of their own. To use a set of tools to discover something new and relevant.

Let me be more specific so you get a clear sense of how this format works.

Lawn chemical mixtures such as Round Up, Garden Safe, or Spectracide are widely used for residential and agricultural use. There is increasing concern on the part of scientists and other citizens about the safety of these chemicals for humans and wildlife. Are honeybees dying because of lawn and agricultural pesticide use? Are human diseases and disabilities such as Parkinson's disease or autism linked to pesticides?

Because many of these pesticides are targeted at nervous system function, this topic not only is relevant to the students' own lives, but also serves as a great focus for a study of neuroscience and behavior for all major levels of biological organization from genetic and cellular to ecosystem and evolution. There are many different chemicals to investigate and lots of opportunities for legitimate new discoveries. At the beginning of the semester, I introduce the students to a variety of pesticides available at lawn and garden stores in our local area. After the students form small research teams, of three to four students each, they choose a pesticide and investigate what is known about its effects on behavior.

The soil nematode *C. elegans* makes for a great study organism. There are lots of behaviors that can be examined, and the neurons and neural circuits that underlie those behaviors are well understood and amenable to microscopic and genetic analysis. After learning a number of behaviors and microscopy techniques with this organism in the workshops, student groups can apply some of those techniques to a specific question they wish to examine. Students read the scientific literature related to their question and learn to develop a feasible experimental approach to test it.

The students decide what behavioral assays they want to use to determine whether a particular lawn or garden chemical affects *C. elegans*. They link their behavioral assay choice to the choice of neuron group to examine. For example, one group might decide to explore the effects of the fungicide Mancozeb. This fungicide has been linked to Parkinson's disease in humans (see Further Reading). Because Parkinson's disease afflicts dopamine neurons, the students examine the effects of either acute exposure or chronic exposure of nematodes to Mancozeb and evaluate a dopamine-mediated behavior, the transition from swimming to crawling (Brody et al., 2012, Further Reading). The group thinks about how to expose the nematodes, at what life stage, and to have carefully considered controls. The students figure out an appropriate sample size and what kind of statistics they need to perform. There are just eight dopamine neurons in nematodes, and there are a few strains of nematodes that are genetically engineered to express the fluorescent protein green fluorescent protein (GFP) within only these cells. The neurons are large and easy to locate using a fluorescent microscope. The student group can put together an exciting project that has never been done before. If they have several weeks to work on their projects, they can examine multiple different conditions or they can revise their experimental design and re-test if they have trouble with their initial plan. The iterative process of design, implementation, analysis, revise, and repeat gives the students an authentic and rich experience with the scientific process of discovery.

This framework can be adapted to different life sciences disciplines such as genetics, cell biology, physiology, and developmental biology. The key features are as follows:

(1) The study organism needs to be easy to maintain and study at more than one level of biological organization. *C. elegans*, *Drosophila*, various plants (unless you are teaching a neuroscience course, obviously!), and zebrafish are all great candidates (see Chapter 4).

(2) There need to be straightforward assays of function or behavior that can be manipulated experimentally within the constraints of the laboratory schedule, the overall semester, and the supplies and equipment at hand.

(3) The laboratory schedule needs to build in time to help students develop feasible experimental plans.

(4) There should be a link between the laboratory projects and the classroom learning. If the laboratory is a separate course, I would suggest that the overall laboratory has a topical focus. By including journal club sessions, the choice of papers you read can link the laboratory and the classroom.

Summary of the Format

The students conduct authentic scientific research. They take a question of interest and develop it into a research project. They evaluate the findings from one experiment, refine their experimental design, and re-examine the question by conducting a new test and building on their earlier results. The quality of the data the groups collect is much improved from having a few weeks to work on the projects. Student presentations reveal that they are deeply engaged and feel a personal ownership toward their projects that often continues past the semester course. The students transform into experts in a small area of the field. Because each project is different, the students listen to each others' presentations with greater focus and often ask questions and offer insights from the perspective of their own area of expertise.

Assessment of Student Learning

What are the main goals of most intermediate-level courses in the life sciences? If we had to name one, it would probably be to develop a depth of knowledge and understanding about a particular subfield, such as Biochemistry or Developmental Biology or Organismal Behavior. How do we achieve depth? Like any subject, a depth of knowledge is best achieved by doing, by frequent practice applying concepts, embellishing the scaffold they constructed at the introductory level. To achieve a depth of knowledge in a foreign language, students practice the language in written and oral form, immersing themselves in the language. To achieve a depth of knowledge in a time period in history, students read many kinds of sources and materials centered on that time period, fleshing out the framework formed by earlier knowledge. In the same way, to achieve a depth of knowledge in a subfield of life science, such as neuroscience for example, students need to immerse themselves in a current area of study and contribute their own discoveries. To do that requires knowledge and background in the field, an ability to identify an unanswered question within that field, and the skill to convert that question into a feasible experimental approach. The format of the intermediate-level neuroscience course described in this chapter accomplishes these goals. How do we assess how well the format achieves these goals?

Goal 1: Achieve a Solid Foundation in the Experimental Approaches to a Variety of Current Research Questions in Neuroscience and Behavior

Both the classroom and the laboratory focus on this learning goal. The quality and depth of student discussion reveals the progress toward this goal. In addition, the quality of the experimental design and the data analysis and interpretation demonstrate progress toward this goal. The student research presentations and follow-up discussion, the quality of student feedback on other group presentations, provide evidence of the level of mastery.

Goal 2: Achieve a Sophisticated Ability to Read and Interpret the Primary Experimental Literature

The classroom assignments and discussions focus on the assessment of this goal. Nonetheless, the way students apply the background literature to support their own projects, as evidenced by their presentations and culminating manuscripts, enables you to assess mastery of this goal.

Goal 3: Formulate a Hypothesis, Design and Conduct a Multilevel Experimental Project Over Several Weeks to Discover New Information About the Relationship Between Genes and Behavior

To assess the students' ability to formulate a hypothesis and design a multilevel experiment that addresses it, have your groups submit a written experimental design in response to a rubric you provide them. Then, in the experimental design consultations with each group, probe individual student understanding and appraise the final design that groups create after the consultation. The rubric can be your guide to student assessment. During the projects, observe the student groups and ask them questions to assess engagement and understanding. The final presentation and manuscripts also provide you with evidence of the degree of achievement of this goal.

Goal 4: Perform and Understand Appropriate Statistical Analysis of Behavioral Data, Gain Confidence in the Use and Limitations of Model Organisms, Computational and Bioinformatics Approaches to Examining Complex Relationships Between Genes and Behavior

The quality of the data analysis, figures, and oral/written presentations of the projects reveal the level of achievement of this goal. Improvement toward this goal is evident when comparing the final figures and presentation of the student group projects with the short assignments from the workshops and skills sessions from the beginning of the term (Appendix A and B).

Goal 5: Become Facile in the "Language" of Neuroscience and Behavior, with a Thorough Mastery of our Chosen Subtopics, as Well as a Keen Ability to Speak and Write on the Discipline

Through class discussion, written assignments, and the research project oral and written assignments, student mastery of this goal can be assessed.

In addition, another important assessment is to poll the students using an end of course survey. Rather than poll about how satisfied they were by the course, which is the most common type of assessment employed, it's important to help them see and reflect upon how much they learned. For this neuroscience course based on reading and integrating the primary experimental literature, a good strategy is to have students re-read a challenging paper from the beginning of the course and then ask them to reflect on their learning experience. An example of the type of questions you might ask is shown in Table 6.2.

As a result of polling a number of classes using this discovery-based format, it is clear that the majority of students believes that their knowledge of the scientific process and their ability to read and understand the primary experimental literature in neuroscience have greatly increased after taking the course (Raley-Susman and Gray, 2010; see Further Reading). Some students are uncertain about whether their content knowledge of all of neuroscience has increased substantially. This was not a goal of the

Table 6.2 Student Self-Assessment Survey

1. When compared with the first time you read a research paper, how much do you think your ability has improved?	5 Greatly improved	4 Somewhat improved	3 Neutral, same as before	2 Not much improved	1 Ability is worse now
2. When reading a paper using a new experimental approach, how much has your confidence improved that you can figure it out?	5 Very confident	4 Somewhat confident	3 No change in confi-dence	2 Somewhat less confident	1 Much less confident
3. My ability to integrate experimental papers with the broader field of neuroscience has been increased by taking this course.	5 Strongly agree	4 Agree Some-what	3 No change	2 Disagree some-what	1 Strongly Disagree
4. My knowledge of neuroscience has increased as a result of taking this course	5 Strongly Agree	4 Agree Some-what	3 No change	2 Disagree Some-what	1 Strongly Disagree
5. My understanding of the scientific process has increased as a result of taking this course.	5 Strongly Agree	4 Agree Some-what	3 No change	2 Disagree Some-what	1 Strongly Disagree

course, however. It is a common misconception on the part of students that taking one course in neuroscience and behavior will enable them to become experts in their content knowledge of this incredibly broad field. It is not the case that content coverage can be equated with learning or mastery of a field of inquiry. Instead, having students hone a set of skills for learning about neuroscience, reading the experimental literature and extracting from it key concepts about nervous system organization and function for example, enables them to delve more deeply into advanced courses in the field. When queried as seniors, most students do feel that the discovery-based course enriched their overall knowledge of the field and their overall confidence in mastering more advanced concepts.

Further Reading

1. Raley-Susman, K.M. and Gray, J.M. (2010) Exploration of gerontogenes in the nervous system: A multi-level neurogenomics laboratory module for an intermediate-level neuroscience and behavior course. Journal of Undergraduate Neuroscience Education (JUNE) 8(2): A108–A115.

2. Some helpful articles about gerontogenes and aging. a.Braeckman, B.P., Houthoofd, K., Vanfleteren, J.R. (2001) Insulin-like signaling, metabolism, stress resistance and aging in Caenorhabditis elegans. Mech. Age Dev. 122: 673–693. b.Taguchi, A., Wartschow, L.M., White, M.F. (2007) Brain IRS2 signaling coordinates life span and nutrient homeostasis. Science 317: 369–372. c.Wolkow, C.A., Kimura, K.D., Lee, M.S., Ruvkun, G. (2000) Regulation of C. elegans life-span by insulin-like signaling in the nervous system. Science 290: 147–150.

3. Some helpful articles about pesticides a.Cicchetti, F., Lapointe, N., Roberge-Tremblay, A., Saint-Pierre, M., Jimenez, L., Ficke, B.W. et al. (2005) Systemic exposure to paraquat and maneb models early Parkinson's disease in young adult rats. Neurobiol. Dis. 20: 360–371. b.Brody, A.H., Chou, E., Gray, J.M., Pokrywka, N.J. and Raley-Susman, K.M. (2013) Mancozeb-induced behavioral deficits precede structural neural degeneration. Neurotoxicol. 34: 74–81.

4. Some useful websites and videos: a.WormBase: www.wormbase.org b.WormAtlas: www.wormatlas.org c.WormBook: www.wormbook.orgd.JoVE: www.jove.com

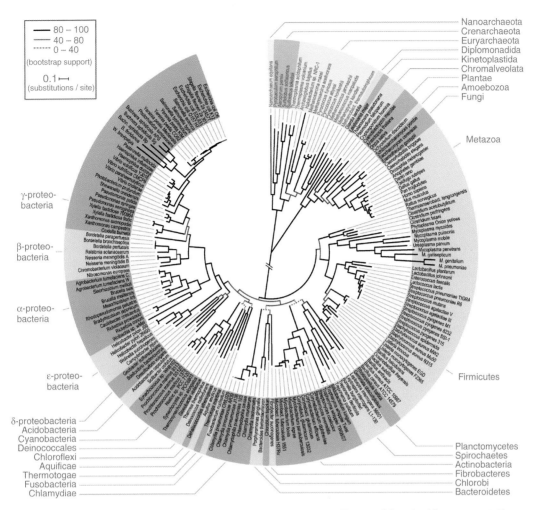

Figure 1.2 **A phylogenetic tree of life no longer really a tree, more like a swirl, as in this representation of evolutionary relationships, based on a genomic study of the rRNA of 3,000 species by David Hillis, Derrick Zwicki and Robin Gutell from the University of Texas. (From Ciccarelli et al. (2006). Reprinted with permission from AAAS.)**

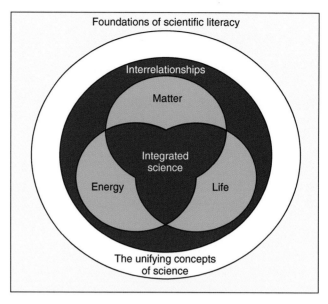

Figure 2.1   Goals for science literacy. (Image reproduced with permission from: http://ae.gov.sk.ca/evergreen/science/part3/portion04.shtml.)

Figure 8.1   The impact of humans on the landscape. (Image from: http://commons.wikimedia.org.)

7 Assessments and Why They Are Important

What is Assessment?

Higher education has joined the Assessment Craze. Federal, State, and Regional organizations want to know if college students are learning and achieving what they are supposed to learn and achieve. Congress wants to know if the money poured into the US Higher Education system is paying off. Do college grads demonstrate "college-level" proficiency and essential skills important for a successful workforce and educated public? Prospective college students and their parents want to know if a particular college or university will prepare them for successful careers. College administrators want to know if the curriculum students explore produces educated citizens with a bright future in the workforce.

The answers to these kinds of questions are often obtained from looking at post-college placements in jobs, in graduate schools, medical and law schools. But do these measures really tell you if your students have learned what was intended? As hard-working faculty, we want to know if our students are learning and are achieving the goals that we set for them at the beginning of the course. If you spend all this time and effort developing and implementing courses and discovery-based laboratories, don't you want to know how well things worked? Maybe you first want to know if changing your way of doing things and incorporating discovery-based laboratories will result in better learning than your current way of doing things. How will you figure this out?

I don't profess to be an expert in the field of assessment. But I am pretty good at figuring out if my students are learning what I want them to learn. This chapter provides some ideas and examples of how you might approach these questions for your own course. There are many great books, articles, and Internet resources that can help you develop

Discovery-Based Learning in the Life Sciences, First Edition. Kathleen M. Susman.
© 2015 John Wiley & Sons, Inc. Published 2015 by John Wiley & Sons, Inc.

assessments for your own course and that can help you learn more about assessment and its importance in teaching and learning. I provide a list of some that I have found helpful at the end of this chapter.

I have three different goals for assessment in my courses.

(1) I want to understand how effectively my students learned the course material and achieved my course objectives.
(2) I want to understand how effectively my course format enabled that learning to take place.
(3) I want to know how well I guided them in their learning as their instructor.

Student Learning Assessments

Mostly, these are just fancy labels for things teachers have been doing for eons. Tests, quizzes, examinations, papers, poster presentations, oral presentations, problem sets … … the stuff we assign for our students to demonstrate they have learned the material, can work with the material, and so forth. These assessments can be direct measures of student learning. How well did they grasp the concepts? How well can they articulate their understanding or their ability to synthesize material? How well can they apply their knowledge to solve a problem? The scores or grades that students earn on these assessments can be used to help guide student learning during the term. In contrast, those final projects and final examinations can tell you to what extent each student achieved your curricular goals.

Assignment-assessments can also be indirect measures of learning. For example, if you have your students prepare a portfolio of different kinds of work in the course, you can have them write a "self-reflection" piece that has them articulate how they approached their learning and how they think their portfolio entry demonstrates their mastery of the material. Another type of indirect assessment that's popular these days is the use of the end-of-class or beginning-of-class quick writing that involves asking students to reflect on an aspect of the reading or class material that they found confusing (e.g., "Your Muddiest Moment") or enlightening. Asking the students to reflect on their own learning and understanding and to tell you about it in writing, so-called meta-cognitive assessment, gives you some key insights into how well they are engaging with and learning the material.

Another great way to assess how well your students are achieving your course goals is to take some class time to look back on a topic or unit in an active way. Traditionally, this is done through examinations, tests, and midterms. But another great way to do this is to create an active assignment or activity that requires students to integrate the course material. Let me give you an example that I use in my intermediate-level and my advanced-level courses in neuroscience and neurobiology (Chapter 6). A key goal for each of these courses is for students to be able to integrate across levels of biological organization. We read and discuss experimental papers that focus on cellular mechanisms, physiological changes, behavioral consequences, and evolutionary changes. Students need to develop the ability to integrate these different ideas into one multilevel understanding of a broader topic (such as learning and memory or stress or neurodegenerative disease). In addition, the students are expected to learn to critically read the experimental literature in the field. To assess the students' abilities, in addition to a midterm type of assignment, I assign each student the task of re-reading an article from earlier in the term. In class, I use the chalkboard to place the name of each article on the board. Each student comes up to the front of the class, provides a one-sentence summary of what he/she re-read, and then draws a line on the board from that paper to a different paper on the board, stating what the connection is. By the end, the board is filled with crisscrossed lines linking the papers together. We build together an integration of the work we have done together and discuss new insights and emergent themes that have arisen from re-reading the work through the conceptual framework that we have built together. This is a powerful way to assess student mastery of course goals. In addition, it makes explicit for the students the learning that they are doing, how their understanding has changed. This assessment also models for them how to take different class material and connect it, synthesize and deepen their overall understanding. You can accompany this kind of exercise with a quick survey at the end of class, such as the one given in Table 7.1.

How do you assess whether a student has achieved the goals of your discovery-based laboratory? Are those goals different from the goals of the entire course? Let's go back to the learning goals for the introductory-level discovery-based course in Biology (Chapter 3).

- How are biological systems organized?
- What is the relationship between structure and function in biological systems?
- What is biological information? How is it stored and exchanged?
- How do biological systems acquire and transform energy?
- How does biological diversity arise?

Table 7.1   Self-Assessment Survey

	5	4	3	2	1
1. When compared with the first time you read this article, how much do you think your ability to read the primary literature has improved?	Greatly improved	Somewhat improved	Neutral-same as before	Not much improved	My ability is worse now
2. When reading a paper using a new experimental approach, how much has your confidence improved that you can figure it out?	Very confident	Somewhat confident	No change in confidence	Somewhat less confident	Much less confident
3. My ability to link a paper to the broader field has increased by taking this course	Strongly agree	Somewhat agree	No change	Disagree somewhat	Strongly disagree

The discovery-based laboratory helps students learn these concepts. In addition, of course, the laboratory portion of the course also focuses on the mastery of particular skills such as using equipment, conducting mathematical or statistical analyses, designing and conducting experiments, and writing and presenting scientific findings. The acquisition of these skills is readily assessed when students prepare a poster or an oral presentation of their work, or when they write about their projects in the form of a scientific manuscript. From the quality of the figures and tables, you can determine how well the students designed, conducted, and analyzed their results. From the writing or presentation, you can discern how well they articulate their understanding of the science they performed.

It's important for me to say a few words about the relationship between grades and assessment. Grades are a crude measure of a student's mastery of the goals of a course or even of a single assignment because they serve more than one purpose. In addition to assessment of learning, they also serve to guide and motivate a student during the learning process. Grades can encourage and motivate students to try harder or to change their learning strategies. They are a form of communication between the professor and the student. Because of these additional roles of grades, it's essential that you clearly articulate for yourself and your students what the learning goals are for each kind of assessment that you create. For each assignment I give my students, I make a grading rubric or other form of guide to how I plan to assess their learning and/or mastery of the material. I spell out my expectations, which are really my learning goals for them, and how I plan to measure their successful achievement of those goals and expectations. I've become a big believer in rubrics and expectation guidelines. They really help me develop assignments that not only measure student achievement of goals, but also help with the actual learning itself.

How does the discovery-based laboratory support the conceptual goals you set for your introductory biology course? This is a tough question that most of us are not used to thinking about much. The majority of the learning that takes place in the laboratory is implicit learning: how to design an experiment; how to conduct and analyze an experiment; how to use statistical tools; and how to present an experiment. The students learn and master these things just by doing, through practice. However, the laboratory also teaches conceptual material, although largely indirectly. Take, for example, the conceptual learning goal of understanding the relationship between structure and function. In Chapter 3, I described a discovery-based laboratory module that focused on this concept. In the Module: *Self-Discovery Explorations of Human Disease Caused by Single Nucleotide Polymorphisms*, students examine the consequences of genetic mutations on cell and tissue function that result in disease. In order to really understand how a SNP in a gene can result in diseases such as cystic fibrosis, students have to be able to relate the structure of the gene to the structure and function of the resulting protein. They have to be able to relate the structure of the protein to the cellular function it governs. They have to be able to understand how those dysfunctional cells affect the tissue or organ system or organism. If the laboratory assignment clearly articulates this conceptual goal in its rubric or guidelines, you can then use this assignment as a way to assess their understanding of the concept. The key in this case is to bring the conceptual goals explicitly into the assignment rubric. If, for example, you have your students give an oral presentation on the basis of their experimental projects, you can have them include an explicit consideration of how the project has furthered their understanding of a particular concept in the field. Or you can have them write a self-reflection essay to include as the cover page to their laboratory write-up that has them articulate how their project has helped them better understand a particular conceptual goal or how their project connects to course concepts. When you include these sorts of components in your laboratory assignment expectations, not only you provide an explicit link between the laboratory and the classroom part of the course, but also you give yourself a way to assess the student learning and mastery of the conceptual goal.

The aforementioned learning goals also reflect a desire to have students develop a depth of knowledge about particular fields of life science. How do we achieve a depth of knowledge? To achieve a depth of knowledge in a foreign language, students practice the language in written and oral form, immersing themselves in the language. To achieve a depth of knowledge about a period of history, students read many kinds of sources and materials centered on that time period. In the same way, to achieve a depth of knowledge in a subfield of life science, such as neuroscience or biochemistry, students need to immerse themselves in the field by reading and doing, through self-discovery and actual discovery. It takes practice, repetition, and frequent exposure.

In addition to these more traditional forms of assessment, you can get some ideas about the students' own perceptions of their learning from surveys. Let's face it. If the students don't think they've learned what you set for them as goals for the course, you have not done your job as the teacher of the course. These *perceived learning* surveys can provide you with an idea of how well you conveyed your goals to your students and how well you helped them to be aware of their own learning. These perceived learning assessments can take a few forms. One is a mid-semester evaluation that is focused on learning styles and learning goals. An example is shown Figure 7.1. The very act of providing feedback and reflecting half-way through the semester involves them actively in their own learning.

Course-Based Assessments

Does your course work? Does the format of your course help your students achieve the curricular goals you have for them? Let's think about this a bit.

Let's say, your course format involves three times a week lectures and a once per week cookbook-style laboratory, that is, the traditional introductory science course format. How do you know this is a successful format? Heck, this is the way science has been taught for decades. Doesn't that mean it must be a successful format? What's the evidence to suggest it's not?

Course surveys and questionnaires, scores on tests? How about the numbers of college graduates with a degree in a life science? How about the numbers of students in your introductory course who go on to take a second course in the life sciences, or perhaps the number who end up majoring in that field? These various measures are all confounded by various unrelated factors and so are rough indicators at best. For example, the particular student population that took your course and are filling out the evaluation survey do not have a basis for comparison. They took your course, with that format. They do not know how much better they could have learned the same material with a different format. They can

Mid-semester evaluation of NEUR201

I am very interested in getting some feedback from you at this point in the semester, so I have a Better sense of your "relationship" with the course and our class community thus far. I will read these Over the break and will endeavour to address any concerns you might have.

So, please take a few moments to provide confidential and anonymous feedback on the following:

1. Do you fell that your ability to read and understand primary literature papers has improved so far this semester? If not, are there things I might do differently to help withs this goal?

2. Do you fell you are learning to take a question of interest and develop a hypothesis that you might be then able to test? *We are developing this skill in class through considering hypotheses that others have tested, and in lab by designing your own experiments.* If you fell you are not learning this skill, are there things I might do differently to help you achieve this course goal?

3. A course goal is to become will-versed in the language of neuroscience to learn to write and speak on the discipline. Do you fell your are beginning to achieve this goal? If not, are there things I might do differently to help you achieve this goal?

4. Do you fell you are learning some important concepts and information about neuroscience and behaviour from multiple different levels of analysis? Do you fell you are beginning to be able to integrate concept across topics?

5. Do you fell that you are on your way towards achieving a solid foundation in the experimental approaches to a variety of research questions in neuroscience and behavior? If not, are there things I might do differently to help you work towards this course goal?

6. Do you fell like we are developing a supportive and engaging class community/community of majors? If not, are there ideas you have about how we might better achieve a good classroom environment?

Figure 7.1   An example of a mid-semester course evaluation.

tell you if they liked the format or if they think the format enhanced their learning. But surveys such as end-of-course evaluation questionnaires do not tell you how your course format contributed to or detracted from student learning.

The test scores are also all wrapped up in your particular course format. You developed the tests from the framework of your course format. If you lectured content, you examined content. If you focused on process, the assignments should focus on process. These kinds of assignments are often not transferable between different styles of course. So, again, you are faced with the problem of not being able to compare across courses.

One common question on student course surveys is, "how likely are you to take another course in this area in the future?" The thinking in this case is that a successful course and format will yield more students interested in continuing to study in the discipline. However, the underlying assumption that the course is the reason for the continued interested is often flawed, particularly for introductory-level life sciences courses. Most students decide to take particular courses on the basis of their ideas of what they want to pursue as a career. For example, many life science majors hope to go to medical school. So, they continue taking courses that will help them get there. They will likely take another life science course after your introductory biology course, whether they liked the format of your course or not.

So, how can you assess how well your course format achieved your curricular goals? The best way is to do an experiment. Teach a few sections of your introductory course the traditional, standard way. Then, teach a few using your different approach, incorporating discovery-based laboratories, for example. Give the same assessment and compare results. Very few faculty and departments can afford to take an experimental approach, but you certainly can assess your existing curriculum before you implement your new one. A few life sciences instructors have done just that. The results? Variable, as we see in the following section.

Example 1: Assessment of Discovery-Based Introductory Biology Course

As described in Chapter 5, we introduced a new format for our introductory biology curriculum in 2004, "Investigations." Originally, we revamped our curriculum in response to student evaluations of our traditional lecture/laboratory two semester introductory biology sequence. Students didn't like the 4-hour weekly laboratories that appeared to "never work" and to be "busy work." The laboratories were grouped into three inquiry-driven modules that were self-contained but that only loosely interacted with the lecture portion of the course. The classroom portion was primarily content-heavy lectures. The first semester stressed evolution/ecology, and the second was cell biology focused. Students tended to identify with one or the other course, as either "cell biologists" or "ecological biologists," and they were reluctant to take courses outside of this narrow focus thereafter. Students felt that to succeed they needed to memorize a bunch of facts and that introductory biology was a "weed-out" course. The faculty felt that it was increasingly impossible to teach a survey of concepts using an encyclopedic textbook. We wanted our students, and ourselves, to be enthusiastic about biology and to see the relevance and importance of all levels of biology in their lives. So, rather than a content-focused traditional lecture/laboratory course, we introduced two courses. One is a topic-focused course without a laboratory. Students learn the fundamental principles of biology by exploring particular topics. Course titles such as "The Life and Times of HIV," "Evolutionary Development," and "Salt, Sex, Salmon and Death" attract students to a multilevel study of biology. The second is a process-oriented, only laboratory course, described in Chapter 5, "Investigations in Biology." The

> 4. Please identify the *C. elegans* on the plate as either wild type or a mutant strain.
>
> 5. You need to make serial dilution to perform a plate dilution assay on a soil sample. You begin with 10 g of soil and add it to 95 ml water for your first dilution (10^{-1}). For tubes with a final volume of 10 ml, how much of your 10^{-1} dilution and how much water do you need to make a 10^{-2} dilution?
>
> How would you create a 10^{-4} dilution sample?

Figure 7.2   Example questions from the Biology 106 Final Practicum. (From Ronsheim et al. (2009). Adapted with permission from The Education Resources Information Center (ERIC).)

course comprises three inquiry-driven and somewhat discovery-based modules, each 4–5 weeks long, where students design and carry out experiments, analyze data, and present their findings in different forms, from poster presentations, to oral presentations, to written manuscripts. In addition to the usual types of graded assessments such as short quizzes, problem sets, and the summative manuscripts or presentations, we conduct a cumulative skills practical final examination. The final examination consists of two parts: a conceptual essay portion that examines mastery of fundamental concepts and a skills practicum. The practicum has about 20 different stations, each of which examines skills and techniques emphasized throughout the semester. Some stations require students to demonstrate their mastery of skills such as using a microscope or balance or spectrophotometer. Some ask students to work with statistics, to solve problems, design a PCR primer, or analyze a gel. Other questions focus on aspects of experimental design or graphical data interpretation. Figure 7.2, taken from the article we published about this course (Ronsheim et al., 2009; see Further Reading), provides a couple of these questions.

What indicators do we have that our discovery-based introductory course has improved engagement and learning of the process of science? In addition to these more traditional graded assessments, we prepared a skills questionnaire that we handed out to our students the first day of the semester and again at the end of the semester (Figure 7.3).

We discovered, not surprisingly, that students rated their knowledge and skill mastery substantially higher after the course, as compared to the first day of the course (Ronsheim et al., 2009). Although I'm sure that most students learn something over the course of a semester of laboratory training regardless of the format, these results were quite satisfying. The most noticeable change was that students no longer self-identified as "skin-in" or "skin-out" but saw themselves as "biologists." They readily thought about biological processes from multiple levels of organization.

Now, 10 years later, we have lots of students taking courses in biology. In fact, we have so many students interested in taking courses in biology that

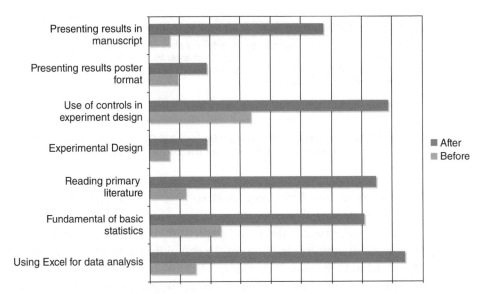

Figure 7.3   **The percentage of students rating themselves as "fluent" in each skill the first day of the semester (Before) and again at the end of the semester (After). (From Ronsheim et al. (2009). Adapted with permission from The Education Resources Information Center (ERIC).)**

we cannot serve them all. Is this the result of the word on the street about our introductory curriculum? What is the word on the street about the course? I recently conducted a survey of a couple of sections of the course. I wanted to know what these students thought about their experience in the course. Did they feel the course prepared them for additional work in biology?

I asked them if this course was a good way to learn introductory biology. More than 60% of the students surveyed agreed. This means that about 40% of them either disagreed or didn't really know. When faculty are asked the same question, the results are similar. Some faculty believe strongly that discovery-based learning is an improvement over content-heavy lectures and traditional technique-focused laboratories. Others believe equally strongly that the loss of content puts students at a disadvantage in upper-level courses. Most are on the fence.

Example 2: Assessment of a Redesigned Introductory Cell Biology Course Using Pretesting and Post-Testing

Laura McEwen and her colleagues at McGill University introduced an inquiry-driven laboratory format to their undergraduate cell biology course and conducted a number of surveys and skills assessment measurements. The overall assessment technique was a series of pre-course surveys followed at the end of the semester by post-test surveys (McEwen et al., 2009; see Further Reading). Disappointingly, the results appeared to suggest only modest gains in conceptual learning, appreciation of science, development of scientific habits of thinking, and achievements of laboratory skills such as experimental design and data analysis. The reason? It appears that the students had really high self-expectations, colored by their prior experiences in high school perhaps and scored very highly in the pretests. The students rated their familiarity with the different skills and concepts high to begin with, so there wasn't much room for improvement. In another experimental approach to assessing the effectiveness of an inquiry-based approach to introductory biology laboratories, Rissing and Cogan (2009; see Further Reading) found significant improvements in student ratings of their mastery of course objectives only when the pretest scores were lower. The students' prior experiences and their expectations going into a course play a huge role in their own perceptions of their learning experiences. Students do not enter into a course as a *tabula rosa*.

Many students report that they are turned off by their traditional introductory courses in life sciences. The cookbook-style laboratory exercises do not get students excited. Sure, students appear to enjoy the laboratories for the first hour or so, but soon they start looking at the clock with increasing frequency. Does a discovery- or inquiry-based laboratory format enhance the student experience?

Yes. Most of the published studies that have re-designed college introductory biology courses report students are generally more satisfied by their inquiry-based experiences than with traditional laboratories, particularly those grouped into multiweek modules with an independent investigation component (Howard and Miskowski, 2005; see Further Reading). We all can appreciate that the more a student enjoys a course or engages with the course the better the learning, right?

It's likely that this increased satisfaction translates into better engagement and hence better learning. If you compare performance across the modules in a discovery-based laboratory course, students perform better on the later modules (Howard and Miskowski, 2005; see Further Reading). This improvement likely reflects a number of things: improved understanding of the course expectations based on instructor feedback and improvements just from the repetition and practice and improved critical thinking/problem-solving skills. Another study by Steven W. Rissing and John G. Cogan (2009; see Further Reading) compared student learning of enzyme function as a result of an inquiry-based laboratory module versus the traditional laboratory exercise. Student participants were all part of the same large-enrollment lecture course, but were in two different laboratory sections. The students who did the inquiry-based module scored better on content questions about enzymes than those in the traditional enzyme laboratory section.

Instructor Quality Assessments

Student evaluation of teaching has been around since the 1920s. By the 1970s, student ratings were firmly entrenched within the fabric of higher education in the United States, and student perceptions of faculty teaching were endorsed by the AAUP as a valid and important part of the faculty promotion and salary process. Thousands of studies have been published about the reliability (or lack thereof), validity, and biases inherent in student evaluation of teaching quality. What characteristics in a teacher are important for successfully managing a discovery-based laboratory experience? I believe the most important characteristic is the ability to trouble-shoot, to think nimbly about the problems that students will encounter when they design and implement their own experiments. The second most important characteristic is the ability to encourage students in the face of their disappointments. Can students reliably evaluate these instructor characteristics?

There's certainly been a lot published on the pros and cons of instructor evaluation questionnaires. Often, these assessments take the form of the end-of-term questionnaire that is a one-size-fits-all bubble form used by many institutions of higher education. This single survey is used by many institutions to evaluate both the effectiveness of a particular course and that of the instructor. It is used for promotion and even salary determinations. Students fill out the surveys most often on the last day of class, before their final examinations, when stress levels are highest. Summative assessment surveys, if they are institution-wide, will ask very general questions that are not tied to the particular curricular or learning goals for any individual course. The following are a few questions you might pose to gauge student perceptions of your role as instructors of their discovery-based course.

- My instructor helped me work through problems and difficulties I encountered during my project.
- When I became frustrated with my project, my instructor provided encouragement and support.
- My instructor provided guidance and support to enable me and my group to work independently on a project of our own design.
- My instructor encouraged me in developing skills in critical thinking and problem solving.

Notice how different these questions are from the "standard" teacher evaluation questions:

- The instructor was organized and clearly communicated course goals.
- The instructor provided helpful feedback.
- The instructor was available outside of class.
- The instructor generated enthusiasm for learning the subject matter.
- The instructor treated students and their contributions with respect.

Interpreting the Data

A very useful resource if you want to learn more about assessment goals and strategies is the publication, "Student Learning Assessment: Options and Resources, published by the Middle States Commission on Higher Education (2007; see Further Reading). I won't even attempt to provide the kind of depth of coverage that they have already done, but I encourage you to read that book and other resources as well. It's clear that assessing student learning and satisfaction is crucial for colleges and universities.

My angle on assessment in this book is to understand the effectiveness of discovery-based laboratories that are incorporated into your courses. How well do these experiences help our students learn about science and encourage our students to appreciate the importance of scientific approaches in understanding the natural world? In general, student satisfaction and self-reported engagement are enhanced by hands-on and inquiry-driven laboratory approaches (Wilke and Straits, 2001; Flannery, 2007; McDaniel et al., 2007; see Further Reading). This just makes intuitive sense. Think back to your own learning experiences. When you are actively working toward solving a problem, or answering a question that you have posed yourself, your experience doing the research work is deeper, more personal. If students get the chance to "do for themselves," they will pay attention more closely, and they will get more from the experience.

The students' prior experiences will color what they expect from your course. Students often feel uncomfortable if a course format is substantively different from their previous expectation or experience with a subject. Many expect that laboratories are exercises that have a right answer. However, discovery-based laboratories don't have correct answers. The process of discovery is messy. Experimental designs turn out to need revision. Doing the experiments may turn out to take longer than initially predicted. The results may not support the initial hypothesis. Maybe nothing pans out. These disappointments and frustrations are a natural part of the scientific process, but are tough to manage and may influence how students learn and how they feel about the field. This minefield of obstacles is tricky for an instructor to navigate. Rather than students thinking, "nothing in bio lab ever works. I hate this field." or "I just can't do science." you want them to think, "My experiment was well-designed. The results indicate that something other that what we predicted is going on. How might I figure it out?"

Assessments, self-reflections, and communications along the way (not just the end of the semester) are crucial ingredients in managing your students' frustrations so that they remain engaged and so that they understand the value of the scientific process.

What to do with the Data?

Every time you teach your discovery-based introductory course, you will have graded and ungraded direct measures of how well your students have achieved your course goals. In addition, you will have indirect evidence about how effective the design of your laboratory is from surveys.

We all want *all* of our students to learn the course material, to achieve the course goals, and to feel strongly engaged and satisfied by the experiences in our courses. There are many factors at play that are unrelated to the course itself. For example, students have differing levels of interest in the subject matter and different motivations for taking a course. In addition, students have competing interests outside of your course, including other courses, employment, extracurricular interests, and personal lives. These competing interests could interfere with a student's mastery of your course goals or with a student's level of engagement.

If we see an improvement in student performance on direct measures of course goals, we will tend to attribute the improvement to the format of the course or to our ability to clearly articulate the goals to the students. If we see an improvement in the survey results, we will tend to believe that the course format was the reason. How much improvement is needed? If the student satisfaction ratings show an increase of 10% or 25% compared to previous curricular designs, will we believe that the course format is an improvement? What's the margin for error in student surveys of this kind? Many of the published articles that present inquiry-based laboratory experiences include some assessment data, most often in the form of pre/post surveys or surveys about scientific skills self-assessment (see Further Reading). Typically, the survey results presented are from one or two different classes of students. The sample size is the number of students surveyed the first time or two that a new course format is implemented.

What happens to student ratings after a "new" course design has been in place for several years, after the newness has worn off?

A key factor in the success of a course is the buy-in of the faculty. Many introductory-level laboratory courses are taught by a large number of faculty members who did not themselves design the course. For larger institutions, these core courses might be taught by adjunct or visiting faculty or even postdoctoral and graduate student researchers. Often, newer faculty members develop their teaching skills by teaching the departmental introductory courses. As a result, instructor expertise and depth of commitment to the introductory courses vary tremendously from semester to semester. In addition, as large, multisection courses get entrenched into a curriculum, they gain inertia and lose vitality, becoming almost rigid and intractable. Discovery-based laboratories reduce that tendency. There is a freshness to each semester because the experiments vary with the different student populations. The implementation of these types of laboratory experiences of necessity retains a certain flexibility and nimbleness, as long as the faculty involved use the format to insert their own interests into the course. Their natural enthusiasm for their own interests will serve to enhance the enthusiasm of their students for the course. Faculty who believe in the value of the course format will convey this positive attitude to their students, who in turn will believe in the value of the course format. And, of course, the student enthusiasm will influence the level of engagement of the students and their ability to learn the course material. These more engaged and enthusiastic students will provide better ratings on course questionnaires. Students will like the course more and so will likely learn more. Image is important. Branding your course in a positive light will improve student learning. If your customers are satisfied, you will have a better teaching experience. A win-win. Keeping tabs on the student satisfaction through surveys will help you to keep your brand popular.

Further Reading

1. Adams, D.A Review of Walvoord, B.E. and Anderson, V.J. (1998) Effective Grading: A Tool for Learning and Assessment. San Francisco: Jossey-Bass.
2. Evidence for effectiveness of active learning. a.D'Avanzo, C. (2013) Post-vision and change: do we know how to change? CBE: LSE 12: 373–382. b.McEwen, L.A., Harris, D.R., Schmid, R.F., Vogel, J., Western, T. and Harrison, P. (2009) Evaluation of the redesign of an undergraduate cell biology course. CBE-LSE 8: 72–78. c.Rissing, S.W. and Cogan, J.G. (2009) Can an inquiry approach improve college student learning in a teaching laboratory? CBE—Life Sciences Education 8: 55–61. d.Howard, D.R. and Miskowski, J.A. (2005) Using a module-based

laboratory to incorporate inquiry into a large cell biology course. Cell Biol. Educ. 4: 249–260.

3. Kloser, M.J., Brownell, S.E., Chiariello, N.R. and Fukami, T. (2011) Integrating teaching and research in undergraduate biology laboratory education. PLoS-Biol. 9: e1001174.

4. Importance of self-reflection and metacognition in learning a.Tanner, K.D. (2012) Promoting student meta-cognition. CBE-LSE 11 b.Fink, L.D. (2007) The power of course design to increase student engagement and learning. AACandU.

5. Sundberg, M.D. (2002) Assessing student learning. Cell Biol. Educ. 1: 11–15.

6. Student Learning Assessment: Options and Resources. Middles States Commission on Higher Learning. (2007) 2nd Edition. Philadelphia, PA.

7. Additional readings on assessment: a. Wilke, R.R. and Straits, W.J. (2001) The effects of discovery learning in a lower-division biology course. Adv. Physiol. Educ. 25: 62–69; b. Flannery, M.C. (2007) Enriching the experience of science. Amer. Biol. Teacher 69: 170; c. McDaniel, C.N., Lister, B.C., Hanna, M.H. and Roy, H. (2007) Increased learning observed in redesigned introductory biology course that employed web-enhanced, interactive pedagogy. Cell Biol. Educ. Life Sci. Educ. 6: 243–249; d. Ronsheim, M.L, Pregnall, A.M., Schwarz, J., Schlessman, M.A. and Raley-Susman, K.M. (2009) Teaching outside the can: A new approach to introductory biology. Bioscene 35: 12–22.

8 Fully Incorporating Vision and Change

The Anthropocene and the Importance of Biology Literacy

We are in the midst of the anthropocene era, a period of changes on a global scale caused directly and indirectly by human activity. Nine billion people by mid-century (Figure 8.1). A warming climate and rising seas. Massive changes in species distribution and diversity. The threat of emerging diseases. Our addiction to fossil fuels and our need to feed ever more hungry mouths have caused earthquakes, felled mountains, diverted rivers to make deserts into farmland, and spewed toxic chemicals into our air, water, and soil all over the planet. Virtually every human activity already is framed in the context of human impact. Every college student needs an understanding of our impacts on the living world around us. A firm grasp of biology in particular and science in general is as essential as reading and writing and communicating. Science must no longer be relegated to a one-semester requirement that students are forced to take to be well-rounded but rather should be valued as highly as the oft-required freshman writing seminar. Can we develop science curricula that produce scientifically literate citizens?

We need a populace with a clear understanding of the anthropocene. If we understand how and why we are now in this rapidly moving geologic era, we can better manage the consequences of our actions by not accepting business as usual. We need business leaders, government officials, doctors, firefighters, politicians, and teachers to understand these issues. Colleges and universities need to put science at the center of their educational missions, *right up there with reading, writing, and critical thinking*. This will require a huge cultural change at all levels of our educational system. Currently, the US K-12 public education system emphasizes reading and writing. Science literacy does not share this imperative, but

Discovery-Based Learning in the Life Sciences, First Edition. Kathleen M. Susman.
© 2015 John Wiley & Sons, Inc. Published 2015 by John Wiley & Sons, Inc.

Figure 8.1   The impact of humans on the landscape. (Image from: http://commons.wikimedia.org.) (*See insert for color representation of this figure.*)

should. Instead, science is an "extra" subject in many places and consistently is crowded out of curricula in favor of other subjects such as leadership training or drug education. Our culture recognizes the essential role of reading and writing, but not of science. Unfortunately, in our culture, science is at the same time valued and hated or feared. Many of our nation's teachers do not feel confident in their science skills because they had only minimal science themselves, and so they unwittingly discourage their students from pursuing science. Biology, in particular the critical role of evolution in biodiversity, is considered controversial; evolution is heretical to some strongly held religious beliefs, and so as a result, many teachers skip over evolution or present an inaccurate version. Money, politics, and power have been poured into preventing the accurate teaching of biology in many states in this country. College instructors have their work cut out for them.

Limited Resources Constrain the Discovery Laboratory for All

This book has focused on how to create laboratory experiences that involve students in the process of discovery, that teach biology and the scientific process through discovery and open-ended investigation. Inquiry-driven and discovery-based experiences enhance student engagement with science. These experiences foster a sense of personal investment that most educators believe is crucial for good learning and a deep appreciation for science. However, as described in Chapter 4, laboratory courses are expensive and resource demanding. Space, supplies, equipment, time, and faculty staffing are all in limited supply. If we believe all students need to learn about biology and these global issues, we need to ask if these approaches scale up. At my college, like at many others, about 1/4 of the first year students take introductory biology, in part to fulfill distributional or other types of college requirements. Most of these students are pursuing careers that involve science or medicine. Even that proportion of students stretches our resources. We do not have the capacity to offer the same experience to all the entering students. Most institutions have similar constraints. Can we modify our discovery-based introduction to life sciences even further to provide this experience for *all* our students?

A related question is whether laboratory-based courses are the only way or even the best way to reach all college students, to engage them on a personal level, and educate them broadly about the imperative to understand biology and how we are affecting the world we live in. Do you need to know how to design an experiment and carry it out in order to be a scientifically literate citizen able to understand the debate about fracking in your home state and to make informed decisions in the voting booth? No. There are alternative approaches that I'd like to consider in this chapter.

Alternative Approaches

Some universities have implemented a different approach – science for nonscientists – as part of "general education requirements." Because a lot of undergraduate students take introductory biology to meet science or quantitative requirements, this course is sometimes the only place to address key issues facing humanity, all of which involve biology. These nonmajors introductory biology courses attempt to foster critical thinking and to provide some introduction to science literacy. Students with an interest in pursuing science do not take these courses, however. Thought to be watered-down and not rigorous science courses, they are considered easy ways to check off a college requirement ("Rocks for Jocks" comes to mind). Often, these courses are large lecture hall affairs with a scientist delivering entertaining lectures to hundreds of students all at once. These courses may unwittingly discourage students from engaging the process of science. The content-heavy lectures deliver a bunch of cool facts. However, introductory biology courses can be a place where all undergraduate students can understand these issues better and so be better informed professionals, policy makers, and citizens. How can life science instructors work toward increasing science literacy while helping students understand the scientific process?

Let's list the characteristics of a scientifically literate citizen and then see if we can adapt our introductory biology course to provide the necessary training.

A science-informed citizen:

(1) Knows how to ask questions about the natural world.
(2) Knows how to approach gaining knowledge to answer those questions through researching what's already known or through experimentation.
(3) Knows how to critically evaluate those sources of knowledge.
(4) Uses this knowledge to make informed decisions, both personal and societal.
(5) Understands and appreciates the value of science to increase personal and societal economic productivity and quality of life.

It turns out that these characteristics are also essential for budding future scientists, who, let's face it, are also science-informed citizens. Being an informed citizen is not only about new discovery, but also about self-discovery. It's about knowing how to ask a good question and how to frame a process for learning the answer. To know how to ask a good question requires some basic knowledge of the governing and major concepts about the natural world, in astronomy, physics, chemistry, biology, geology, and the climate. Understanding these concepts is a key part of being an educated person. Every citizen should have some understanding of natural laws such as the conservation of energy, the nature of chemical bonds, the evolution of living organisms, the characteristics of climate, and the nature of infection and immunity.

Incorporating these concepts into every college student's education requires an institutional change in academic mission. You can make a start with your introductory biology course.

Envisioning Introductory Biology for the Science-Literate Citizen

Biological systems are major players in climate change, the human food supply, energy consumption, and human health. All of the key political, economic, and societal systems interact with biological systems, making an understanding of biology crucial for all educated citizens. Biology and life sciences programs in higher education have an obligation to provide learning experiences for all students. Some key curricular goals are as follows:

- Increase engagement and interest in science, particularly the biological sciences
- Increase an appreciation for how science is done and how new knowledge and understanding of biological systems are acquired
- Increase the appreciation for the importance of scientific ways of thinking in understanding the natural world
- Foster a curiosity and lifelong quest for knowledge about the natural world.

Introductory Life Sciences: The Discovery-Based Classroom

Once you have gained some confidence in developing and implementing discovery-based laboratory modules that give students a more sustained engagement with the process of biology, you will be ready to fully embrace a discovery-based approach to your entire course, both classroom and laboratory, or even to develop a discovery-based course that does not have a laboratory. By emphasizing the published experimental science literature, many of the learning goals for science literacy can be accomplished without a discovery-based laboratory component. You could offer instead a course that meets every weekday for an hour or so in a classroom setting of 20–40 students. Students, in preparation for reading selected experimental papers, could view online tutorials and videos or could conduct virtual laboratories. Students could explore Internet resources, virtually visit researcher laboratories, look up related articles to contextualize, and bring to life the world of that experimental paper. These resources are more dynamic and engaging and would give students key background context for reading the assigned experimental paper. Then, in the classroom, students could discuss the articles, could work in small groups to design experiments, to integrate their knowledge together, or to conduct library-based research to deepen their learning and their understanding.

Sure, there's a lot of value to hands-on laboratory work, and budding scientists and science professionals need that experience. However, only a subset of introductory life sciences students plans on careers in science. Imagining yourself doing an experiment, planning an experiment without actually doing it, and thinking through observations and data using a scientific process even outside of a laboratory can also be rewarding and can deepen and enhance learning. Think about how much you can learn from watching a documentary on the History Channel or the Discovery Channel. Those shows are basically lectures, interviews, and talking heads. The audience (in the comfort of their living rooms) sits passively and absorbs the images and information. Why are they so memorable? Why do we learn from these documentaries even if we are not engaging in active learning or hands-on inquiry? Because these are stories, images, characters, and plots. Because we see, hear, experience, and imagine. We want to be part of the action, part of the drama unfolding before us. Incorporating ways to get your students mentally active, whether it be in a classroom or a laboratory, is the lynchpin to a successful learning environment. We can develop classroom experiences that bring in these crucial elements of story-telling and documentary film and active involvement. Videos, speakers, student panels, investigative projects, and student discussion, can all bring life to any scientific discipline (Table 8.1).

If you want your introductory life sciences courses to include laboratory work as an entry to a biological science major, but you do not have the resources for every college student to have that laboratory experience, you can offer accompanying laboratory sections that the subset of potential science majors would take. The "classroom" portion would be designed identically and would be taken by all students. For some discovery-based classroom projects (see the following section), you could consider forming student teams where some of the students are working in the laboratory and some are not, but the groups collaborate on planning experiments, doing the literature-based research and even analyzing the results.

Students can learn and understand how science and scientists work; they can learn how to question and search for answers to those questions, without having to do the science. Because, honestly, being a scientist is more about asking the questions and figuring out how to go about getting

Table 8.1   Strategies to Engage Undergraduates in Introductory Biology

Idea	Some Sources
1. Videos on Climate Change, Human Impacts	(a) nationalgeographic.com (A Way Forward: Facing Climate Change) (b) National Geographic: The Human Impact (youtube.com)
2. Scientist Interviews- videos and podcasts	(a) The Institute for Genomic Biology (www.igb.illinois.edu/video) (b) National Institute of Environmental Health Sciences Interview Series (www.niehs.nih.gov/news/video) (c) Science Friday Series (sciencefriday.com)
3. Laboratory Websites	Some examples of good ones: (a) Tuna Research and Conservation Center (Stanford University. www.tunaresearch.org/research) (b) HHMI News (www.hhmi.org/news)

the answers than it is about operating equipment. Good scientists spend most of their time reading about what work has been done by others, refining and rephrasing their questions and figuring out approaches than they actually spend time at the laboratory bench. They are actively searching out knowledge and applying it to questions they find relevant and intriguing.

Organizing the Discovery-Based Classroom: An Introductory Life Science Course for All Students

Ensure that all undergraduates develop the level of biological literacy they need to understand, contribute to, and make informed decisions about the complex problems facing the world. (from Vision and Change, 2011)

For this introductory life science course, I have organized the topics into units to align them with possible discovery-based laboratory modules. These classroom topics can be taught without a laboratory if resources are too limited to provide laboratory experiences for all students. However, importantly, all students learn the same material. This would be a common point of entry to the biology or life sciences major, not a "nonmajors" course at the periphery of a curriculum. Introductory biology may be the only means to acquire a basic level of science literacy for all those students not planning on a career in science, including the future teachers of our next generation, so a lot is riding on introductory biology. Two key goals for all students are to learn how to integrate facts into a larger conceptual context and how to pursue a question of interest on one's own. The skills emphasized in this course are essential for all educated citizens, scientists and nonscientists alike.

Unit One: Food and Energy

All life depends on the energy from the sun to transform and exchange matter among living and nonliving systems on Earth. Energy transformations and exchanges occur at all levels of biological organization and link all organisms into complex and dynamic living systems. Plants, primary producers, transform the energy from the sun into organic carbon bond energy through photosynthesis. All organisms convert the organic molecules into chemical energy in the form of ATP in the process of respiration to power all the activities of cells and tissues. Energy and matter flow and are transformed and cycle through all levels of organization,

Table 8.2   Skills and Concepts Addressed by Food and Energy

Guiding Questions	Concepts	Skills in an Accompanying Laboratory
Why does life on Earth depend on the energy from the sun?	Carbon; molecular diversity of life	
How is the sun's energy transformed into the matter of living things?	Metabolism: photosynthesis and respiration	Experimental design: Enzyme assay (if lab) or Virtual enzyme lab
Are there other sources of energy utilized by living organisms?	Energy Transformations: conservation of energy, mass	Interpretation of quantitative data, problem solving, in-class group work, case studies
How do fossil fuels promote global warming? Are there fuels that don't?	Global Energy Cycles: nutrient cycling	Student discovery projects
How do organisms obtain food (fuel)?	Food: agriculture, soil biology, soil chemistry Acquiring food: animals and plants Energy allocation, nutrition and digestion	Student discovery-based projects: literature research, integration, critical thinking Student discussion; primary literature on evolution
How do we feed the human population?	Population density Population ecology Population dynamics	Mathematical modeling and quantitative reasoning, interpretations of quantitative data

through living and nonliving interconnected systems. Students can discover how energy cycles operate at these different levels by considering a number of important issues facing us. A great way to organize this topic is to identify key guiding questions (Table 8.2) and ideas of interest through an initial brainstorming session.

Involving your students to identify questions and ideas of interest to encourage curiosity and sustained engagement

Students need a voice; they need to be active partners in your course. Some areas that will likely be of interest to all your students include learning about invasive species, or learning about how human agricultural systems are organized to provide food for billions of humans. In understanding these two questions, students need to understand what food is, how energy is needed to produce food, and how food produces energy. They need to know how energy and matter are related and how energy flows through global systems. They need to understand the relationship between photosynthesis and respiration, as well as the cycling of nutrients between living and nonliving parts of ecosystems. Table 8.2 outlines the guiding questions, concepts, and skills for this unit.

Each concept is introduced as a guiding question or two that organizes the content into a conceptual framework. The questions are broad enough to enable individual curiosity and to provoke exploration and discovery. Let's take, for example, the first few questions from the table.

How does life on Earth depend on the energy from the sun? How is the Sun's energy transformed into the matter of living things? Are there other sources of energy utilized by living organisms?

The matter that is life exists in the form of chemical molecules that interact and recombine to both form the structural elements and the energy substrates that fuel the structural formations. Emphasizing the transformations and the flow of energy through molecular, cellular, organismal, and global cycles can underscore the interconnections among organisms such as microbes, plants, fungi, and animals. Two central processes to understand are photosynthesis and respiration, both ancient and complex biochemical pathways that involve spatial and temporal assemblies of protein-based enzymes. In answering these initial questions, students combine thermodynamics, bioenergetics, and enzymology with evolution and structure–function relationships and systems biology. Rather than lectures, design assignments that require student discovery and investigation. If your course includes a laboratory, this unit is a great place for the photosynthesis discovery laboratory described in Chapter 3.

The Importance of Enzymes

There are a number of enzyme function tutorials and virtual laboratories that could be assigned as homework or projects to give students some experience with how enzymes are studied (see Further Reading). These exercises or projects could be made into classroom discovery experiences by linking them with student-generated explorations of enzymes important in the digestion of particular foods or compounds. For example, if the class is examining how agricultural systems go about providing food to sustain billions of humans, you could have students investigate some of the key enzymes involved in food production, such as those involved in the dairy industry, in the making of cheese or lactose-free milk, for example.

How Do Organisms Obtain Fuel? How Do We Feed the Human Population?

Here address the notion of food webs and interconnections. Where do humans fit into these webs? How does agriculture work? How is agriculture changing the equation? The relationship between agriculture and human evolution is an active area of investigation, with numerous primary experimental papers (see Further Reading) that students can read and explore. For example, students can examine studies about the evolution of the lactase enzyme in humans and the correlation of that with domestication of milk-producing livestock. Bioinformatics approaches, such as those described in Chapters 3 and 5, would deepen the connections between evolution, genes, and behavior. Students can look at how major commercial pesticides work and the evolution of resistance to them. These relevant and current topics emphasize the relationships between fuel, energy, food, and evolution from a multilevel perspective.

How Do Fossil Fuels Promote Climate Change? Are There Fuels That Don't?

We are deluged with political and advertising claims of "clean coal" and "climate-saving" natural gas through hydraulic fracturing technology (fracking). In this case, you can compare biological energy transformations with the human uses of fuels and energy to power human societies worldwide. Including these questions in your unit will reinforce the fundamental principles of energy transformations, energy balance, and global cycling. Students could research and explore the chemistry and biology behind how we use fuels, different types of fuels. Students can compare the efficiency of biological fuel use, such as respiration, metabolic pathways, with human use of fuels of various sorts.

Unit Two: Climate Change and Other Human Impacts

The greatest challenges facing us and our next generations are the complex consequences of our industrial revolution and exploding population. Humans have altered every ecosystem on the planet through changes in climate, rising seas, chemical contaminants, and fragmentation of habitats. The anthropocene era in which we find ourselves marks the most rapid and severe mass extinction that the planet has yet faced. A firm grasp of the rate of change and the rate of recovery are key aspects to understanding human impacts. So, for example, how quickly can a species evolve or adapt in other ways in response to a change in climate? Which species can adapt and which cannot? How do scientists compute mutation rates, evolution rates, biodiversity, and rates of climate change? Table 8.3 shows the guiding questions, concepts, and curricular goals for this unit.

These concepts and their underlying biology can be dramatized by having student groups select an ecosystem or biome to focus on for a discovery

Table 8.3   Skills and Concepts Addressed by Climate Change and Other Human Impacts

Guiding Questions	Concepts	Skills and Goals
How do biotic and abiotic components of ecosystems interact?	Ecosystems: abiotic and biotic interactions	Student research group work: identifying a question of interest, creating a hypothesis for a literature-based discovery project
How have climate change and habitat change, both as a result of human activities, affected organisms in different biomes?	Climate Change, Habitat Change: Biome and Organism Effects	Reading the experimental literature; Developing ideas and organizing an argument
How might organisms adapt physiologically to anthropogenic changes to ecosystems? Are some organisms more resilient than others? Why?	Physiological adaptations: Sensory systems, Hormonal systems, Epigenetics	Critical thinking and writing, synthesis, integration
What types of behavioral changes might contribute to a resiliency towards rapid anthropogenic ecosystem change? Are some organisms more behaviorally adaptable than others? Why?	Behavioral adaptations: migration, dispersal, biomes, life history	Presenting and communicating scientific concepts
How do organisms adapt evolutionarily to these changes? Why, ultimately, is evolutionary level change important? Is it essential? Why or why not?	Evolutionary Adaptations: natural selection, mutations, inheritance, meiosis, epigenetics	Integrating across levels of biological organization

project, to be the class experts about: oceans, deserts, arctic, rainforest, for example. By consulting the experimental literature and their textbook resources, students can investigate the changes that have occurred within their ecosystem as a result of human activity. They can discover the biodiversity within their ecosystem and identify key species for the health of that system. For that organism, they can examine the current work on the evolution, genetics, development, or physiology and how human activities have affected its life history. If that organism has not been well studied, the groups could identify a model organism that could provide insights to these issues.

Your student groups could actively engage the experimental literature and could interview, via Skype or phone calls, scientists conducting the work they are reading about. They could also view videos and online tutorials about the methodology and experimental approaches the scientists use. Presentations or a research symposium with poster presentations could be a culminating assignment for this module. If you have a laboratory component to your course, this curricular unit could encompass a laboratory-based discovery experience similar to the cyanogenic clover module described in Chapter 5. If the course has an associated, optional laboratory that is taken by a subset of students, groups of laboratory and nonlaboratory students could develop joint projects where laboratory students contribute their experimental work to a fuller literature-based project conducted by the entire student group. In addition, this curricular unit works well with a unit on the effects of pesticides or other human-induced changes on primary productivity, such as described in Chapter 3 using *Elodea* or some other photosynthetic model system.

Unit Three: Health and Disease

A third major issue we are all concerned about is human health and disease. Major new advances in medicine and epidemiology involve new molecular biological and genomic technologies. All educated citizens should have a good understanding of the biological principles underlying the identification, treatment, and spread of disease. As with the other units, a great way to begin this section of the course is to brainstorm with the students about what aspects are of interest to them to explore and discover. What questions do they have? Table 8.4 shows how some questions might organize fundamental concepts, skills, and goals.

Students can choose a disease or health condition that they are personally motivated to explore. Many or perhaps most of the ideas the students will be interested in will involve a knowledge and understanding of genetics, principles of cell division, reproduction, inheritance, and development. The spread of disease and its prevention also involve some working knowledge of the immune system as well as behavior. Furthermore, this topic weaves back to an understanding of evolution and natural selection. Class time can be spent in guided exercises working with concepts of genes, alleles, mutations, gene regulation, and inheritance. Students can then apply those principles in their individual investigations of genetic aspects of the disease or condition they are researching. Linking cellular and molecular aspects of their condition to the tissue and organ level will involve exploration of concepts of mitosis, meiosis, tissues/organ development, and the relationship between genes and protein function. How cells work within tissues in health and disease to cause organ-level dysfunction will deepen their understanding of structure–function relationships and will also introduce students to the

Table 8.4 Skills and Concepts Addressed by Health and Disease

Guiding Questions	Concepts	Skills
How does the immune system fight infection?	Immune System function Health Status and Reproduction	Literature-based discovery research Critical thinking and reading
What are the relationships between microorganisms and their larger hosts (e.g., the microbiome)?	Microbes and infections – humans as ecosystems Procaryotes and eucaryotes	Integration with earlier sections of the course
How do infections occur?	Membranes and organelles	
Why is antibiotic resistance an emergent health problem? How does it inform us about evolutionary processes?	Antibiotics and evolution of resistance Fungi, nematodes and other types of infectious agents	This can link with discovery-based lab, or videos/tutorial lab
Why is cancer such a widespread issue for human health but not for many other organisms?	Cancer: cell growth regulation, cell signalling Mutations, DNA repair, regulation of genes Cell cycle regulation	Cell growth measurement: tutorial, quantitative reasoning
Why is cancer a model for evolutionary change?	Cancer as evolution	

myriad roles of the immune system. There are a number of videos and tutorials online about immune system function and aspects of immune regulation that will enhance learning about this complex topic (see Further Reading). If you have an accompanying laboratory portion for the course, the module described in Chapter 3 about antibiotic resistance can be tapped to accompany discussions in class about epidemiology, spread of disease, and evolutionary-level thinking about many diseases such as cancer, influenza, and disease resistance.

Students can present their multiweek, discovery projects in a full class poster session or two. Alternatively, you could have students give mini lectures about their disease or prepare a public service announcement video. These projects can encompass multiple levels of biological organization and can touch upon ideas and themes from earlier parts of the course.

Summary of This Chapter

By the end of this discovery-based classroom experience, students have an individualized understanding of key biological concepts and the scientific process relevant to their lives and interests. Not all students will have worked with the same particular content. For example, one student might have a deep understanding about genetically modified organisms (GMOs) in agriculture, the effects of climate change on ocean systems, and the development of chemotherapy resistance in lung cancer, while another student might have examined the effects of pesticides in the development of resistant pests, the impact of human-generated noise on birds and antibiotic-resistant microbes. Yet, all the students develop a keen appreciation for how biological science is done and a deeper understanding of major concepts in biology such as evolution, the relationship of structure to function, energy transformation, and information flow. Through the discovery process encouraged by this course, students would have practiced many of the skills we value, such as critical thinking, communicating science to different audiences, quantitative reasoning and modeling, and how to take a question of interest and develop it into a feasible scientific study.

Many college and university life sciences programs do not have the resources to offer discovery-based laboratory courses at the introductory level to all students. It almost goes without saying (but I will say it anyway) that discovery-based laboratory experiences dramatically enhance the learning and engagement with science. However, biology courses that are discovery-based in the classroom can be deeply fulfilling and meaningful without a laboratory. It's similar in a way to the appreciation you can get for art and art history even if you yourself do not put paint to paintbrush.

Laboratory and hands-on experiences in science are crucial early on, in elementary, middle, and high school, and remain crucial for those students wanting to pursue careers in the life sciences. However, laboratory experiences are neither essential for learning the core concepts of biology, nor for developing a life-long appreciation for science. I believe students can learn the process of scientific discovery and can become scientifically literate through discovery-based classroom experiences.

Combining Science Literacy Training with Science Career Training

Many colleges and universities offer separate, nonlaboratory, non-majors courses for students fulfilling general education requirements and a separate, laboratory-based, sequence of introductory courses for those students considering careers in science or science-related fields such as medicine. The nonmajors courses are sometimes thought to be "dumbed-down" and get reputations such as "Rocks for Jocks." These kinds of attitudes contribute to the reluctance students have to take non-majors courses. On the other hand, when nonscience majors are required to take an introductory science course that is geared for those few students who will end up becoming scientists, the desire to take those courses is also seriously hampered by negative attitudes. The discovery-based course described in this chapter could in fact serve two audiences, two purposes. Let's imagine a first-year course of study that is not laboratory based, which emphasizes reading the primary literature, developing questions into experimental ideas, and learning how to think about and communicate about biological science. Science literacy, basically. For students interested in majoring in biology or pursuing medical degrees, have separate but accompanying laboratory courses that dove-tail with the classroom experiences at the introductory level. These could even be half-semester courses focused on a single discovery-based module. Then, the intermediate-level laboratory courses for majors, such as that described in Chapter 6, organized around discovery-based projects, could be modeled on how graduate students are trained.

Concluding Thoughts

Discovery, whether it be at the laboratory bench, in the field, or in the library stacks, is a key component to learning. The powerful feelings that accompany discovery fuel deeper connections with a discipline and drive a hunger for more discovery. There is a tremendous sense of satisfaction in figuring something out for yourself. Tapping into those powerful and deeply personal connections is at the center of the ideas generated in this book.

Practice is another key component to learning. Rote memorization is not an effective strategy for learning concepts such as evolution or gene regulation. Rather, effective practice involves working with the material, applying it to solving problems, and deliberately manipulating material in new ways. Some repetition is essential, perhaps better called rehearsal, and involves attentively and deliberately revisiting the material in an intentioned way, with the purpose of improving performance. Learning science is no different from learning a language or a musical instrument in this regard.

Motivation is perhaps the single most important component to good learning. We've all seen this in our college courses. Some students are extremely motivated to learn the material. They show up for class prepared, ask questions, and come to office hours. These students do well academically as well. What motivates them? Some are motivated by a strong desire to go to medical school. They are motivated by the possibility of earning an A. So, the desire to perform well academically is the motivator. These students may perform well on assignments but may find that they do not retain the information, or perhaps they have only a superficial understanding of the concepts because their focus was on the right answer to the assignments. Some are motivated by a strong interest in the subject matter. These are the students who do more than what you ask for on assignments. They delve more deeply into their laboratory projects. They read in the discipline on their own, for fun. They ask questions in class that are more integrative, more big picture.

Every student will have some area of personal interest that relates to biology and life science. A disease they are interested in because a family member or friend suffers from it. A worry about global warming or climate change. An interest in a particular kind of organism or region of the world. Any of these interests can be used to motivate students to learn more. The discovery-based laboratories and course ideas presented in this book will help you to create learning experiences that tap into personal motivation and interest. Personalizing the course as much as you can while still keeping unified themes and approaches will ignite the motivation that many students need in order for good learning to take place.

Inspire your students with your own story. Why are teachers important in good learning? Sure, teachers organize the material and guide students through modeling how to solve problems. However, crucially, teachers motivate students to discover within themselves a lasting interest through inspiring them. Your enthusiasm for your subject is contagious. Use it. You can also include videos or Skype interviews with inspirational leading scientists whose work you and your students read as part of the course. Consider exposing your students to inspirational individuals who have made use of their interest in science in nonscience arenas, to show your students the utility of a science background in life after college.

Further Reading

1. www.phschool.com/science/biology_place/labbench/lab2 This is a virtual enzyme lab that introduces students to concepts in enzyme catalysis, kinetics and measurement.
2. Harper, K.N. and Armelagos, G.J. (2013) Genomics, the origins of agriculture, and our changing microbe-scape: time to revisit some old tales and tell some new ones. Am. J. Phys. Anthr. 57: 135–152.

3. Malmstrom, H., Linderholm, A., Liden, K., Stora, J., Molnar, P., Holmlund, G., Jakobsson, M. and Gotherstrom, A. (2010) High frequency of lactose intolerance in a prehistoric hunter-gatherer population in northern Europe. BMC Evol. Biol. 10: 89

4. Types of immune responses: innate and adaptive. https://www.khanacademy.org/science/biology/immunology

Appendix A: Laboratory Instructions for Behavioral Experiments Using *Caenorhabditis elegans*

The following are the instructions I hand out for the two behavior workshops.

Caenorhabditis elegans, a soil nematode, has become a powerful model organism for studying the genetic and molecular bases of behaviors ranging from locomotion to reproduction, feeding, egg-laying, chemotaxis, and others. In addition, this organism is fast becoming established as an appropriate model organism for evaluating the molecular mechanisms of sensory neurobiology and various forms of neural plasticity. Intriguingly, despite a tremendous wealth of information about cell circuitry and development, the areas of sensation, behavior, and behavioral plasticity are still developing for this organism. The good news is that there are many good studies waiting to be performed!

In this 2-week workshop, you will become familiar with the locomotory, feeding, chemotaxis, and mechanosensory behaviors of *C. elegans* (perhaps more familiar than you ever wanted to become!). We will focus on the fundamentals of behavioral experimental design, statistics, and close, microscopic observation. This workshop will introduce you to the art and science of designing behavioral-level experiments with *C. elegans*, as well as the statistical tests commonly used. As a laboratory group, you will conduct two different types of behavioral assay, one to probe mechanosensory processes and the other to probe chemosensory processes. You will need to consider sample size as well as the kind of statistical test.

Discovery-Based Learning in the Life Sciences, First Edition. Kathleen M. Susman.
© 2015 John Wiley & Sons, Inc. Published 2015 by John Wiley & Sons, Inc.

Learning Goals and Expectations

Goals for this module are as follows:

(1) *Design and carry out two behavioral-level experiments* using worms with a genetic defect affecting each sensory behavior and wild-type worms. You will design and carry out an experiment comparing the worm strains.

(2) *Analyze your two data sets using two-sample comparison statistics.* You will need to consider sample size, variability, and inter-rater reliability. Your data will serve as the starting point for your independent project, so the quality of these data is important.

(3) *Prepare*, for the Workshop 1 Assignment, a well-designed and clearly labeled figure with figure legend of each behavioral experiment. This assignment is due as a hardcopy two-part figure *at the beginning of Laboratory*.

Part 1: Initial Behavioral Observations of Wild-Type and Mutant Worms

Spend a few minutes becoming familiar with the behavior and appearance of *C. elegans*. You will need to practice picking worms to new plates (this is a skill that is initially tricky to master!), as well as moving larger quantities of worms using a liquid method. Expect to be somewhat frustrated here, but do not give up. Jot down your observations of the behavior of wild-type and mutant worms in the presence and absence of their food source, *Escherichia coli*. You will observe two strains of worms:

Wild-type worms (N2 strain)
mec-4 mutants (CB3274) – these worms have defects in mechanosensation.

Workshop 1A: Mechanosensory Behavior Experiments and Statistical Analysis

As a laboratory group, design and conduct a simple experiment, with appropriate controls, exploring mechanosensory behavior.

You will design a *one*- or *two*-variable experiment (in other words, statistics no more sophisticated than one-way ANOVA). During your planning session with the instructor, your design will be evaluated and modified as needed. In addition, you can use the computers to go to the WormBase website (Google "WormBase"), and you can look up what's known about the mutant we have available, *mec-4* (CB3274).

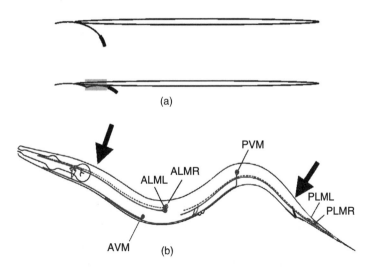

Figure A.1   *Using an eyebrow hair to test gentle touch sensitivity.* **(a) Positioning (top panel) and gluing (bottom panel) the eyebrow hair to the tooth pick. The thickened black line indicates the shaft of the hair; the gray area indicates the location of the glue. (b) Animals should be touched by stroking the hair across the body at the positions of the arrows. The six touch receptor neurons are indicated. (This figure is from www.wormbook.org)**

We have the following additional supplies and materials for mechanosensory behavior assays:

- Rubber bands for plate tap assays or eyebrow hair "brushes" for gentle touch assays
- Worm picks for moving worms
- Plates without food

The following are some common assay techniques, based on the chapter in WormMethods by Anne Hart (http://www.wormbook.org/chapters/www_behavior/behavior.html)

Stroking with an Eyebrow Hair

The initial and most generally used method to test for gentle touch sensitivity is to stroke the animals with an eyebrow hair that has been glued to the end of a toothpick (Figure A.1 from WormBook Methods).

The hair is sterilized by dipping it into a 70% ethanol solution and drying by shaking (don't flame the hair). Sometimes, bacteria accumulate on the hair; they can be removed by poking the hair into the agar on a spare plate.

Animals are touched by stroking the hair across the body just behind the pharynx (for the anterior touch response) or just before the anus (for the posterior touch response; see Figure A.1). Actually, touching the animals in any position along the touch receptor processes will generate a response. Touching the animals near the middle of the body (near the vulva in adult hermaphrodites), however, yields ambiguous results because both the anterior and posterior touch circuits can be activated. The animals should not be poked with the end of the hair, as this provides a stronger stimulus and can sometimes evoke a response, even in touch-insensitive animals. Similarly, animals should not be touched with the end of a platinum worm pick. Animals should not be touched at either the tip of the nose or the tip of the tail, as even animals lacking the six touch-sensing cells often respond.

Routinely, animals are said to be touch sensitive if they respond to stroking with the eyebrow hair by stopping movement toward the hair or by moving away from the hair (sometimes, the touch stops moving animals without having them reverse their direction of movement). Touch-insensitive (Mec) animals fail to respond to the hair but do respond to prodding with a worm pick. A partial response is one in which the animals move away from only some of the touches.

Plate-Tap Assay

Wild-type animals will move (adults usually reverse direction) in response to their plate being tapped. This stimulus often happens when plates are placed firmly on the stage of a dissecting microscope. Touch-insensitive animals do not respond to this tapping. Although not as accurate a measure of touch sensitivity as touching with an eyebrow hair, this is a rapid assay that has been used to screen for touch-insensitive mutants. Place a rubber band around a plate containing a small population of 30-50 worms. Then, while viewing the plate of worms under the microscope, pull the rubber band out a measured distance and snap it back onto the plate. Quickly count the number of animals whose movement reversed or stopped within a 2-s interval. Record as percentage of animals on the plate.

Statistical Analysis

You and your partners will conduct the independent experiments that you designed. You should collect all your measurements and use the provided computers to begin your data analysis. You and your group should decide how best to represent your data in figure or table form. Your goal is to create the best, most informative, and most concise visual presentation of the data, with appropriate labeling and descriptions in legends, axes, and so on. The software program, VassarStats, is a user-friendly and freely available package that I encourage you to use for the statistical analysis.

The statistical background you had in Biology 106 or Psych 200 is sufficient for the tests you will need to perform on your data sets.

For general data acquisition and calculations, I recommend Microsoft Excel. For statistical analysis, you can copy/paste columns of data into most statistical packages (especially VassarStats). You can use your spreadsheet as a place to accumulate your data throughout the semester. You can also create pretty good graphs, charts, tables, and so on using Excel. Its statistical abilities are okay, but the stats module is not included with the base Excel package.

You will need to calculate means, standard deviations, and standard error of the mean for your data for good figure or table presentation.

Workshop 1B: Chemosensory Behavioral Experiment and Statistical Analysis

This workshop will provide training and practice designing and carrying out a chemotaxis assay. Chemotaxis behavior is crucial for most animals (and even other motile organisms such as bacteria and protists), allowing them to detect and move toward food and to find potential mates, by responding to chemical cues that are both volatile and water-soluble. In response to the chemical cues, animals will move toward the chemical source, moving up a gradient of the chemical. Chemotaxis is also an important behavior for detecting and responding to harmful compounds in the environment, allowing animals to move away or be repelled by potentially toxic chemicals or chemicals emitted by other organisms to ward off predators.

C. elegans has a sophisticated chemosensory system involving 11 pairs of specialized chemosensory neurons. They can detect numerous different compounds, can respond with movement toward or away from concentration gradients of the compounds, and can learn both nonassociatively (habituation) and associatively (avoidance or attraction when paired with food or starvation) to chemicals. There are many different types of chemotaxis assays. For your independent projects, consult the chapter in WormBook (http://www.wormbook.org/chapters/www_behavior/behavior.html) and other sources in the scientific literature. Our assay is based on the JoVE paper noted in the syllabus but has been modified by me for our laboratory environment.

Chemotaxis Behavior Analysis
Chemotaxis supplies, considerations:

- Odorant (10% butanone in 95% ethanol) and vehicle (ethanol)

- Petri plates with chemotaxis medium – without *E. coli* bacteria (food source) or sodium chloride (a potent attractant in low concentrations and repellant in high concentrations)
- 1M sodium azide (paralyzes the worms for ease of counting)
- 95% EtOH
- Timer
- Micropipettors, microfuge tubes, and tips for moving worms
- Sharpies.

The following is a rough sketch of a common chemotaxis test set up, based on the JoVE paper (but using smaller, 6-cm plates):

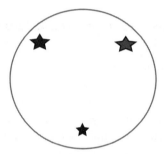

The odorant spots should be placed at equal distances from the worm spot and as far apart from each other as possible, but not at the very edge of the plate. Use a ruler to make measurements.

Label your plates on the bottom with dots that represent the chemoattractant (10% butanone) and the control (ethanol) and the spot where you plan to place the worms. The first thing you will need to do is to set up your chemotaxis gradients. First place a 1-µl drop of 1M sodium azide at each spot on your plate about 15 min before your assay. This will paralyze the worms so that you can better count them at the end of the assay. Then, overlay 3 µl of your attractant at one spot and 3 µl of ethanol at the other. Place the lid on and set the plates aside until just before your assay. *Be sure not to open the lid because this will alter the gradient that is forming.* Then, you will collect worms from the plates into labeled microfuge tubes using a micropipettor. Flood the plate with 1 ml of distilled water, swirl gently and suck up the liquid (with worms) into the pipet, and place them in the tube. Let the worms settle by gravity for about 5 min. While the worms are settling, add an additional 1 µl of chemicals to the chemotaxis plates to refresh and enhance the gradients (quickly replacing the lids).

Then, remove most of the liquid on top of the worms, being careful not to suck up the worms, and discard. Using a fresh pipette tip, add 250–500 µl of distilled water to each tube and gently resuspend the worms by flicking on the side of the tube. Depending on your experimental design, dispense small amounts (5 μl or so) of the worms onto the spots on the chemotaxis assay plates using a cut yellow pipette tip. Either wick away the extra fluid to release the worms as shown in the JoVE video or wait for the drop to evaporate before beginning to time the assay.

At the end of the timed assay, count the worms at each spot and also find the total number of worms on the plate. Compute the chemotaxis index as described in the paper.

$$CI = \frac{[(n_{Butanone}) - (n_{EtOH})]}{[(Total - n_{Origin})]}$$

Remember that each plate is one number. You will need to have at a minimum three plates per treatment or condition in order to perform statistics.

According to the JoVE paper you read in preparation for this laboratory, the worms, when starved for 1 hour, are quite motivated to move toward various chemoattractants such as butanone. For this experiment, you can either examine various pre-experiment or experimental conditions or compare mutant with wild-type navigation in the chemotaxis assay. In planning your experiments, be sure to keep things simple enough that you can complete the experiment with an adequate sample size *and* begin the statistical analysis during the laboratory period. I would recommend you avoid complex two-factor designs. Feel free to ask me if you have questions while designing your experiments. *Do* not plan to test things such as learning or other modifications of chemotactic behavior.

Appendix B: Instructions for Microscopy Workshop

A major advantage of *Caenorhabditis elegans* for cellular neurobiology is that the nematode is transparent, making microscopic imaging possible with simple wet mounts on slides. In addition, the ability to make transgenic animals expressing green fluorescent protein (GFP) or TexasRed under the control of promoters for virtually any gene has truly revolutionized the field (the impact of this technological advance was made clear by the awarding of the Nobel prize to Dr. Martin Chalfie). The *C. elegans* research community has a growing selection of nematode strains that express fluorescent proteins that fluoresce at various wavelengths. A consortium of nematode biologists is creating GFP strains for every gene in the *C. elegans* genome, almost free of charge to the worm community.

GFP is one of a number of proteins that fluoresces in response to different wavelengths of light, producing the bioluminescence of a number of coelenterates, including the jellyfish *Aequorea aequorea*, found in the waters of the Atlantic and the Mediterranean, and the sea cactus, *Cavernularia*. Another closely related compound, aequorin, was discovered to fluoresce differently depending on the concentration of calcium to which it is exposed. That compound formed the basis of a series of calcium-sensing fluorescent dyes used to measure small changes in calcium in living organisms in response to different cell signals. In addition to these biologically relevant and naturally occurring compounds, scientists have created related compounds that fluoresce at many other wavelengths, to allow for double or triple labeling to look at subcellular localization. Some of these include TexasRed, RFP (red fluorescent protein), YFP (yellow fluorescent protein), chameleon, cherryRed, and many more.

The use of fluorescent compounds to visualize the location or the timing of events in living organisms has revolutionized cell biological research. With the cloning and sequencing of the GFP gene, GFP is used to follow

Discovery-Based Learning in the Life Sciences, First Edition. Kathleen M. Susman.
© 2015 John Wiley & Sons, Inc. Published 2015 by John Wiley & Sons, Inc.

patterns of gene expression in living organisms, and the fusion of GFP to other proteins allows researchers to monitor the whereabouts of a particular protein in a single cell. The technique is tremendously important for neuroscientists to be able to study particular subtypes of cells.

We will use several of these strains to study the neuroanatomical features of different types of neurons during the workshop. Our goals are to become familiar with the fluorescent microscopes, cameras, and imaging software in the Biology department's Microscopy suite. In addition, we will learn how to make figures, labeled and with calibration bars for scale, in order to be able to prepare publication quality figures. The final goal of this workshop is for you to become familiar with the normal (wild-type) appearance of different neuronal subtypes in *C. elegans* so that you can then assess possible effects of exposure to common lawn chemicals.

Because of space constraints in the Microscopy suite, we will split into two groups: one will come to laboratory from 1:30 to 3:30 pm, and the other will come from 3:30 to 5:30 pm. You will work in pairs to prepare wet mounts of different worm strains for viewing with the microscopes.

Assignment for Workshop 2

Prepare a publication quality figure that compares the different subtypes of neurons available to you (motor neurons, ventral nerve cord neurons, mechanosensory neurons [these are glutamatergic] and dopaminergic neurons). Your figure will have multiple panels and should be labeled clearly. Important features should be noted with arrowheads or other labeling devices. In addition, your figure should include a scale bar. The figure should have an explanatory legend, complete with title. Attached to the figure you will write an explanatory "results" paragraph. This paragraph is analogous to the text in a Results section of a paper that would accompany the figure. Your entire submission should be two pages: the first being the actual figure with legend, and the second being the explanatory paragraph. This assignment may be submitted electronically as an email attachment to me because of the color requirements of the figure. *Please* have your name in the filename for the attachment!

Procedure for Preparing Wet Mounts of C. elegans

Simply making a wet mount and adding a coverslip will squash the nematodes, so first you need to make an agar or agarose pad to cushion them. This will not completely eliminate but will drastically reduce crushing the worms. The pads perform best when they are fresh, so first you will make yourself several slides with agarose pads.

Prepare three slides as follows:

(1) Make two slides with a piece of tape on them on only one side of the slide (Figure B.1). The tape will provide a uniform thickness for the agar drop to form a level pad. Place a clean slide in between them on a flat surface. Using a Pasteur pipette, place a small drop of melted agar (from the 65°C heat block) onto the center of the clean, middle slide.

(2) Cover the drop by placing another clean slide on top, perpendicular to the slide with the drop. Press gently so the agar drop is flattened to a circle that is the thickness of the labeling tape. Avoid air bubbles.

(3) Allow the agar to solidify (takes about a minute or two). Gently pull apart the slides. The agar circle will stick to one of the slides. Rest the slide, agar side up, on the bench top and allow it to dry for a few minutes.

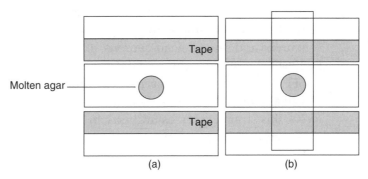

Figure B.1   Diagram of agarose pad preparation. (a) Initial preparation. (b) Place a clean slide on top of the molten agar droplet to form a flat pad.

Mounting live animals

(1) Place a 5-µl drop of M9 buffer (in a rack at room temp.) containing 1 M sodium azide (NaN_3) onto the center of the agar pad. The azide will anesthetize the worms so that they will not move.

(2) Transfer a few animals to the drop using a worm pick. Be sure to use sterile technique! Do not touch the agar pad with the worm pick or scoop up pieces of agar.

(3) Grasp a cover slip on one edge with a pair of forceps. Place the opposite side of the cover slip onto the microscope slide, near to your drop of agar. You may want to use a pencil to keep this edge from slipping as you lower the other side of the cover slip. Gently lower the cover slip onto the drop of water, allowing bubbles to slide out as you lower it. Your wet-mount is now ready.

(4) Transfer it to the fluorescent microscope and observe your worms, first under low power and phase contrast, then under 40× power with phase contrast, the fluorescent illumination. Try to distinguish between adults and larvae.

We will be looking at three different strains of worms, all of wild-type physiology and behavior, but that are expressing GFP (or TexasRed) under different neuronal promoters.

OH7547: this strain is expressing TexasRed under a pan-neuronal promoter, so all neurons will be red in the fluorescent scope. In addition, worms are expressing GFP under the control of a promoter that governs dopamine synthesis enzymes. So, the dopamine neurons will be green. With the confocal microscope, which allows two-laser imaging, the dopamine neurons are yellow because they express both TexasRed and GFP.

LX929: this strain is expressing GFP under the promoter for the enzyme involved in ACh production (choline acetyltransferase). Thus, all motor neurons and several head neurons, express GFP.

BC12648: this strain expresses GFP under a cytoskeletal protein promoter. GFP is found in mechanosensory neurons, which are glutamatergic, as well as a number of neurons in the ventral nerve cord.

PVX4: this strain expresses mCherry in all neurons. It's simply gorgeous! This strain is just for fun.

Make a wet mount of several worms of each type (a separate, labeled slide for each strain). Use the camera imaging set up to gather several images, at low magnification and up to 40×. Make a folder for yourself and collect your images into the folder. Be sure to label the image files in a way that is useful for selecting the images you want to use for your figure. In addition to those images, you should also observe the worms closely and take notes. We will not be using the confocal microscope for this workshop, so you will notice that it's difficult to get all the neurons in focus at the same time. Thus, you may need to select only the most relevant comparative images.

Index

Discovery-Based Learning in the Life Sciences, First Edition. Kathleen M. Susman.
© 2015 John Wiley & Sons, Inc. Published 2015 by John Wiley & Sons, Inc.